SCHRÖDINGER'S MECHANICS

World Scientific Lecture Notes in Physics, Vol. 28

SCHRÖDINGER'S MECHANICS

by
David B. Cook
University of Sheffield
England

World Scientific
Singapore • New Jersey • London • Hong Kong

QC
174.26
W28
C 657
1988

Published by

World Scientific Publishing Co. Pte. Ltd.
P O Box 128, Farrer Road, Singapore 9128

USA office: World Scientific Publishing Co., Inc.
687 Hartwell Street, Teaneck, NJ 07666, USA

UK office: World Scientific Publishing Co. Pte. Ltd.
73 Lynton Mead, Totteridge, London N20 8DH, England

Library of Congress Cataloging-in-Publication Data

Cook, David B.
    Schrödinger's mechanics.
1. Wave mechanics.   2. Schrödinger equation.   I. Title.
QC174.26.W28C657   1988   530.1'24   88-33779
ISBN 9971-50-760-9

SCHRÖDINGER'S MECHANICS

ISBN 9971-50-760-9
ISSN   0218-026X

Printed in Singapore by JBW Printers & Binders Pte. Ltd.

for I. P. C.

Preface

Among the wealth of early mediaeval documents still extant is a collection
of letters which were exchanged between two cathedral scholars of the early
years of the eleventh century — Ralph of Liege and Reginbald of Cologne
— concerning their geometrical studies. Both of these scholars had copies
of Boethius'commentaries on parts of Aristotle's works which were made at
the very end of the Roman Empire of the West in the sixth century. They
were corresponding about Boethius'exposition of the familiar theorem that
the sum of the interior angles of a triangle is two right angles. They both
knew the theorem and could follow the proof but it is evident from the
correspondence that niether of them knew what the interior angles of a
triangle were!

This nightmarish situation shows that the tendency to over-abstraction
and a false economy of exposition ("elegance") is by no means new. The
feelings of these learned men of nearly a thousand years ago must find an
echo in the modern student when he picks up many modern discourses
on quantum mechanics; one can follow the proofs and derivations without
too much difficulty but one is often at a loss to know what the theory is
about. What are the referents of the symbolism? To be able to follow a
logical proof line by line, knowing the definition of every term, and yet not
to know what is being said is both the strength and weakness of modern
mathematical exposition. Its strength is its generality, its weakness is its
use when the formalism is not simply a skeleton but an interpreted theory.

The weakness of logic is that it is *only* logic; a machine for the transformation of statements. There is nothing *illogical* about the assumption
that, since no-one has observed them, dinosaurs never existed or that only
recently has the far side of the moon come into existence. These assumptions are not illogical, they are merely preposterous; yet their substance has
been made the foundation of a whole interpretation of quantum mechanics
(the positivist view).

It is quite common these days to see claims in the popular scientific
literature that modern physics is on the verge of generating a single theory

which will "explain everything". Of course, the meaning of "explain "is rather subjective but to me (in a physical context) it means, at least, a knowledge of the laws of interaction between the entities in the theory and an understanding of the structure and interpretation of the dynamical laws governing the interactions. Modern theories are, almost invariably, semi-empirical in that the formalism is carried through and finally calibrated against known results — the laws of interaction are not known. There are still prolonged and heated discussions about the meaning of quantum theory — it is still only accepted as a working formalism. This kind of approach is strikingly exemplified in a popular article by F. Dyson

> The student begins by learning the tricks of the trade ... then
> he begins to worry because he does not know what he is doing.
> Then quite unexpectedly the student says to himself "I under-
> stand quantum mechanics"or rather he says "I understand now
> that there isn't anything to understand". (Scientific American
> Sept. 1958)

quoted in fascinated horror by Alfred Landé.

In this work, which is a preliminary version of a more systematic approach, I take the "classical"view of the subject matter of physical theories; that is, that quantum theory is a theory of the structure and dynamics of the very light and the very small. Like all theories its area of applicability is limited and, again like all scientific theories, its structure and concepts evolved from those of preceding and "neighbouring"theories; in this case classical particle dynamics.

Taking this view, no problems of interpretation are encountered except the ones which occur in *every* physical theory: "why does the dynamical law take the particular form which it does"and "how does this law relate to 'higher'and 'lower'theories"? The most debilitating effect of the influential positivist view of quantum mechanics is the dictum that there can be no theory "below"probabilistic quantum theory. In the development given here some prominence is given to this problem since it is clear that such a theory must be possible; alas I can offer no clue to its form!

Historically, mathematics has arisen from contact and interaction with real physical problems (*pace* Bourbaki!) and mathematicians have purified the rough-hewn products of physical scientists. Perhaps the role of mathematics in physical science can be inferred by analogy from a quotation from another early manuscript; this time Pope Gregory the Great's advice to St. Augustine in his mission to the English (who were, then as now, old-fashioned and refractory) in 601 AD:

> The idols are to be destroyed, but the temples themselves are
> to be aspersed with holy water, altars set up in them and relics

> deposited there. For if these temples are well-built, they must be purified from the worship of demons and dedicated to the service of the true God. In this way, we hope that the people, seeing that their temples are not destroyed, may abandon their error and, flocking more readily to their accustomed resorts, may come to know and adore the true God.

Pope Gregory was offering advice about how best to use existing church buildings to attract the English to the new christianity, not how best to make mathematics acceptable to physical scientists but his statesmanlike message is clear.

This presentation will, no doubt, contain many blemishes, from spelling slips to mathematical evasions; all of them are entirely my fault and not all of them are avoidable. But the largest omission is of a bibliography which, I hope, will be provided in a more final version of this work. However, I can scarely avoid mentioning the influence of two of the most penetrating writers on the subject matter of this work: Professor M. Kline and Professor L. Mayants.

D. B. Cook (September 16, 1988)

Contents

SCHRÖDINGER'S MECHANICS

Chapter 1

General Orientation

1.1 Introduction

Quantum mechanics continues to be the subject of disputes which flare
up from time to time and never seem to be any closer to a satisfactory
resolution and, increasingly, the disputants seem to be split into mutually
hostile camps with their own independent interpretation and independent
literatures with little area of common ground. However, all agree on the
mathematical and manipulational aspects of quantum theory; it is the interpretation of the symbolism which is at the root of the disagreements.

Historically, quantum theory is in a unique position among the theories
of physics. Unlike most major theories it was not preceeded by a series
of incomplete, intuitively-derived theories but it arrived on the scene in
a more-or-less final mathematical form but without a satisfactory physical
interpretation of the symbolism. In both of the original formulations — the
matrix mechanics and the wave mechanics — it was possible to calculate
the allowed energies of simple physical systems in almost perfect agreement
with experiment and yet not have any idea what the non-energy quantities
in the symbolism actually stood for. This fact has coloured the thinking of
several generations of physicists and, in combination with a now discredited
philosophical movement of the early years of this century, has engendered a
distinct line of approach to the whole of physical theory: "what is important
is the numerical calculation of experimental measurements not intuitive
pictures of reality".

Attempts to give the mathematical formalism a physical interpretation
— particularly in the case of wave mechanics — did follow immediately
on the discovery of the key equations. But confusions about probabilities

and about the interpretation of some quantities in classical mechanics (angular momentum, the Hamilton-Jacobi equation) led to the development of conflicting "schools". These confusions came to be crystallised into a series of paradoxes; some of which are simply the result of a confusion about the interpretation of probabilities ("Schrödinger's Cat", "Wigner's Friend", "contraction of the wave packet", "role of the observer"). The other, more subtle, group of paradoxes stem from an inadequate understanding of the relationship between angular momentum and the homogeneity of space compounded with the relationship between individual statistical results and probability predictions (Einstein-Podolsky-Rosen paradox, Bell's Theorem). In some cases, the *scope* of quantum mechanics is misunderstood; the idea that quantum mechanics, like all theories, has a more or less well-defined area of applicability is temporarily forgotten.

The solutions of these paradoxes are all known and in theoretical physics literature but they do not seem to have found their way into theoretical physics texts and so the confusions continue indeed even flourish! The emphasis on seeking a satisfactory interpretation of quantum theory or the emphasis on ignoring the need for such has, unfortunately, tended to obscure the real difficulties associated with quantum theory: the search for a theory which underlies the probabilistic approach of quantum theory. Indeed there exists a "proof" that no such theory can exist. This proof is performed within the confines of quantum theory! The fact that such a theorem can be accepted as genuine is astonishing particularly since this proof is formulated in a rigorous axiom-system for quantum theory: quantum theoretical axioms and mathematics can only generate theorems about quantum mechanics and nothing else. In particular it cannot generate theorems which may contradict parts of quantum mechanics in special cases.

In this work it is hoped to disentangle mathematics from physical interpretation and, in particular, to try to present the relationship between the intuitive, scientific development of a theory and the axiom systems which arise once the theory reached a mature form. The treatment is not historical but methodological (and therefore quasi-historical or, as the critic might retort, pseudo-historical) and we shall attempt to show how axioms may be chosen naturally as the "fall-out" from a single central idea which itself is rooted in the mainstream of theoretical mechanics. We shall not be concerned with the development of an axiom-system for quantum mechanics for reasons which will become clear as the work unfolds but perhaps a word or two about the relationship of axioms to the physical theory are in order here.

1.2 Physical Theories and Axioms

The point of view taken here is that quantum mechanics is a theory about molecular, atomic and sub-atomic matter; its energies and distributions in space and time. Thus, treating quantum theory as part of the main body of theoretical physical science, we hope to develop a theory of a small part of reality (the realm of the very small and light) which will be mathematically expressed but intuitively accessible. Like all other physical theories its scope of applicability is, presumably, limited. Any theory of physical reality will be rich enough to have formal structures "embedded" in it. These formal structures can be abstracted and treated by conventional mathematical methods and any mathematical system which models part of reality is sure to have a degree of complexity. But these same mathematical structures may also be abstracted from other, quite unrelated, physical theories. That is, by co-incidence, the same mathematical structure may be found in (abstracted from) more than one physical theory. This result may be suggestive of an inner relationship between the theories or may be merely that the two theories share a common "skeleton". The applications of Poisson's equation to heat flow and elasticity provide a familiar example of such a "co-incidence".

Clearly, one cannot decide on the basis of the mathematical structure what the meaning of the symbols are. Meaning is provided, not by the formal skeleton, but by the substantive axioms of the physical theory; that is, by the physical interpretation of the mathematical structure. In particular the appearance of a partial differential equation second order in spatial variables does not necessarily mean that any real (physical) waves are involved in the underlying physical theory.

If, therefore, it is found desirable to formulate a physical theory in an axiomatic fashion such an axiomatisation can only be considered satisfactory if it provides a full physical interpretation of the symbols involved: the primitive (undefined) terms must be *described* and not left simply undefined. In science it is not so important to be able to define something as it is to know what it is; to be able to describe it in physically acceptable terms.[1]

The sought-after physical interpretation is always to be found in the actual historical development of the theory and its relationship to other theories which partially overlap in subject matter and provide common

---

[1] This point is made admirably by Dickens in "Hard Times" where the unfortunate Cissy is asked at Gradrind's school to "define a horse". Cissy, whose whole life has been spent with horses and knows more about them than Gradrind and all his school put together, cannot define a horse. The definition is provided on demand by the obsequious Bitzer, who has probably never been near a *real* horse in his life.

concepts or, at least, common limiting concepts.

So, to return to our starting point, since the formal mathematical structure of quantum mechanics is known and some of its relationships to "neighbouring"theories are known, the most promising way to a satisfactory physical interpretation must be an examination of the way in which quantum mechanics developed from other physical theories whose physical interpretation is known. To risk a repetition: the physical interpretation of quantum theory cannot be had from an examination of its formal structures because those formal structures are not unique to quantum theory.

1.3   Prerequisites

We are not attempting an rigorous axiomatic formulation of quantum theory here as we have emphasised in the last section but it is, nevertheless, useful to collect together at this point those techniques and theories which we shall silently assume during the course of the work.

We make the usual assumptions associated with physical science: the subject-matter of our study (molecular, atomic and sub-atomic particles) exist independently of us and they exist in three-dimensional Euclidean space E^3 (we shall not be concerned, except in passing, with relativistic theories). The properties of E^3 will be assumed to be modelled by R^3 (the product of three copies of the real number system) and all the usual properties of such a model of real space will be assumed: co-ordinate systems, manifolds etc. etc. Real space is distinguished from conceptual, mathematical spaces (configuration space, phase space, momentum space) by just the attribute of existence in the ordinary sense of the word. This takes us naturally into a consideration of mathematics.

The whole of "classical"mathematics will be used as necessary without comment; occasionally set-theoretic *notation* will be used where precision is necessary but set theory will not be needed. The use of mathematics in this way is, of course, completely standard but does carry with it the assumptions of ordinary classical logic which does seem to exclude the use of more exotic logics in quantum theory since all interpretations of quantum theory agree on the use of the mathematical apparatus.

There will be the need from time to time to use the results of physical theories where these bear on the behaviour of the referents of quantum theory (sub-atomic particles). When this happens we shall again silently assume the validity and utility of the theory. The main theory used in this way is, of course, a theory which provides the laws of force which act on micro particles: electrodynamics. In so far as we are concerned with the quantum mechanics of particles (not fields) we shall assume the validity of

Maxwells equations and the familiar law for the force acting an a moving charged body. Classical particle mechanics is a special case since we shall initially assume its validity and later deny it!

All physical theories attempt to describe and explain the self-generated structure and development of objectively existing systems which occur whether or not the systems are observed or subject to measurement. Measurements on and observations of systems are performed to obtain data about the structure and behaviour of the system which may have relevence to a theory of its autonomous behaviour. Any theory may well generate descriptions of behaviour which are not accessible to observation or measurement (theories of stellar evolution or geological rock formation, for example) or which are only capable of examination by indirect means using technologically complex apparatus and long chains of reasoning involving theories other than the one being tested.

The relationship between theory and experiment is not a simple one and, although unequivocal experimental results must always be the final deciding factor, it is very rare for single measurements to be decisive. The reasons for this are obvious enough. If we take, for simplicity, the measurement of a quantity which is predicted by theory to be a real number, the results of a systematic set of measurements will, typically, be a pair of rational numbers: an estimate of the quantity and an error bound. The result of a single measurement will, typically be a subset of the real numbers (within which the value should lie): an element of the standard topology of the real numbers. In symbols, if S is the system and T the theoretical quantity the theoretical result is:

$$T : S \times U \to R$$

where U is a system of units and R is the real numbers. However, the experimental magnitude (T') is something like:

$$T' : S \times U \times M \times M' \times E \times O \times \ldots \to P(R)$$

where M is the set of methods of measuring T, M' is the set of theories about how these methods work, E is the set of equipment used, O the set of experimental workers etc. etc. and $P(R)$ is the set of intervals in the real numbers (with rational end points). Clearly T and T' are different mappings with different domains and different ranges and the relationship between them is not simple.

As we have said, the ultimate arbiter of theory is observation and experiment but any theory which hopes to model part of reality will be richer than its set of numerical predictions: it should tell us what is happening to generate those numbers. And conversely, the measurement of any physical quantity brings to bear theories and methods other than those involving the object or process under investigation.

Chapter 2

Classical Mechanics

2.1 Introduction

Classical particle mechanics occupies a position in modern theoretical physics which is at once ephemeral and essential: formally non-existent but actually indispensable. In a strict formal sense quantum mechanics can be axiomatised in such a way that it is completely independent of classical mechanics and this kind of approach is useful for certain investigations about quantum theory as a theory in contrast to quantum theory as a description of part of reality. But quantum mechanics grew out of classical particle mechanics and the conceptual structure and physical interpretation of quantum mechanics grew out of the concepts and referents of classical mechanics. To treat quantum mechanics as actually independent of classical mechanics in such an a-historical way is rather like treating India as independent of the U.K. For formal (political) purposes India is, of course, independent of the U.K. (and any other state) but its independence from the U.K. is different, for example, from its independence from Iceland. India has *become* independent of the U.K. but it has always been independent of Iceland. Any attempt to understand the structure, culture and institutions of the Indian nation cannot be made without reference to the role of the British Raj. Similar considerations apply to the relationship between quantum mechanics and classical particle mechanics. The formal independence of quantum mechanics from classical particle mechanics is different from its independence from (e.g.) thermodynamics. Quantum theory has developed from classical particle mechanics and has *become* independent of it whereas it was always independent of classical thermodynamics. It simply is not possible to understand quantum mechanics without some familiarity with the

structure and concepts of classical particle mechanics in the latter's more
general forms.

However, if we may push the analogy a little further, there are instances
when formal independence is all-important and decisive. These are generally in what one might term in both cases in "legal" matters. In matters
of the here-and-now the Indian state is sovereign and no appeal to history
should sway the legislature. Similarly, in formal questions about the inter-
nal structure of quantum theory the well-chosen axiom system is decisive.

But we are concerned here with the evolution of structure *and* concepts
and so we will review the main features of classical particle mechanics. For
practical purposes, that is to say for the purposes of actually solving problems in classical particle mechanics, the equations of Lagrange are the end
of the development of classical mechanics. If the trajectories of particles
are what is required — $\dot{q}(t)$ and $q(t)$ for motion in a particular field of force
and for given initial conditions — then the most usual procedure to set up
and solve Lagrange's equations. However, in investigating the structure of
classical mechanics Hamilton, Poisson, Routh and Jacobi made discoveries
which, although of little practical value at the time in the calculation of
particle trajectories or any numerical predictions, were of enormous help in
clearing the way for the development of much of modern theoretical physics.
It is precisely those latter developments, which define mathematical quantities which are quite remote from the original concepts of Newton, which
are crucial to an understanding of quantum mechanics.[1]

2.2   Lagrange's Equations

In modern terminology Newton's famous equation relating force and acceleration is a vector equation. Lagrange's development of this equation
changes this emphasis in two ways:

1. It replaces references to force and acceleration by references to two
   (scalar) energy functions. Forces are replaced by "generalised forces" which
   are derivatives of a single scalar function (for conservative systems)
   and accelerations are replaced by derivatives of the kinetic energy

---

[1] It is a matter of some historical interest to note what a stickler for intuitive accept-
ability Hamilton was in all his work. He was one of the last prominent mathematicians
to accept the existence of negative numbers because he thought them unreal and he was
held up in his work on quaternions (spinors) for many years by difficulties of interpretation. "And, while agreeing with those who had contended that negatives and imaginaries
were not properly *quantities* at all, I still felt dissatisfied with any view which should not
give to them, form the outset, a clear interpretation and *meaning*" Preface to Lectures
on Quaternions (1853)

function. Thus, in Lagrangian mechanics the emphasis is placed on relationships between derivatives of energy functions.

2. The equations which result from application of the ideas in (1) are expressed in arbitrary co-ordinates. This has the effect of making the general theory free from "accidental" factors due to particular co-ordinate systems since intuition is not a reliable guide to (e.g.) the kinetic energy expression in prolate spheroidal co-ordinates. In practical applications of Lagrange's equations the ability to choose any co-ordinate system often enables a particularly appropriate choice of co-ordinates to be made which simplifies the integration of the differential equations.

The whole thrust of the development is to replace vector equations by scalar equations and concentrate the essence of the dynamics into a few (actually one) scalar functions from which the forces and accelerations may be obtained by differentiation. We proceed by starting in Cartesian co-ordinates, transforming to general co-ordinates and comparing the final expressions with the original Cartesian expressions for similarities of structure and for new intuitions for the further development of mechanics.

The motion of a system of N particles is governed by $3N$ Newtonian equations of motion which, in Cartesian co-ordinates, takes the form

$$F_i = \frac{d}{dt}(m_i \dot{x}^i) = m_i a_i \qquad (2.2.1)$$

where F_i is a force component acting on a typical particle (mass m_i)[2] The velocity components dx^i/dt are written $\dot{x}^i$ as usual. The kinetic energy of such a system of particles is just the familiar expression.

$$T = \frac{1}{2} \sum_{i=1}^{3N} m_i (\dot{x}^i)^2 \qquad (2.2.2)$$

Now

$$m_i \dot{x}^i = \frac{\partial T}{\partial \dot{x}^i} \qquad (2.2.3)$$

and so 2.2.1 becomes

$$F_i = \frac{d}{dt} \frac{\partial T}{\partial \dot{x}^i} \qquad (2.2.4)$$

---

[2] Actually, there are only N m_i not $3N$ and so the m_i should not have the same suffix as the x_i. But the m_i will shortly disappear, in the meantime: $m_1 = m_2 = m_3 : m_4 = m_5 = m_6 :$ etc.

If we confine attention to laws of interaction between the particles which are conservative then there is a potential energy function

$$V(x^i) = V(x^1, x^2, \ldots x^{3N})$$

whose gradients are the force components

$$F_i = -\frac{\partial V}{\partial x^i} \tag{2.2.5}$$

and, in this case, (4) becomes

$$\frac{d}{dt}\left(\frac{\partial T}{\partial \dot{x}^i}\right) = -\frac{\partial V}{\partial x^i} \tag{2.2.6}$$

which is the desired result in Cartesian co-ordinates in the sense that it is set of scalar equations ($3N$ of them, one for each i) and refers only to (derivatives of) the energy functions $T(x^i)$, $V(x^i)$.

Now 2.2.6 depends on the use of 2.2.2 the kinetic energy T; in Cartesian co-ordinates the expression for the kinetic energy is very simple and, in particular, depends only on the $\dot{x}^i$ not on the x^i. This is not true for general co-ordinates, not even for all orthogonal co-ordinates and much of the mechanics which follows depends on being able to choose co-ordinates at will either for general, theoretical purposes or for particular applications.

Let us therefore introduce a general co-ordinate system for our system of N particles as $q^i (i = 1, 3N)$:

$$q^i = q^i(x^1, x^2, \ldots, t) \tag{2.2.7}$$

i.e. $q^i = q^i(x^j)$ for $i, j = 1, 2, \ldots, 3N$

We require that the q^i be sufficient to describe the motions of the particles by insisting that 2.2.7 is invertible, that is we may recover the x^j from the q^i by functions of the form

$$x^j = x^j(q^i, t) \tag{2.2.8}$$

We may now express the equations and expressions leading up to 2.2.1 in terms of the q^i by elementary transformations. Since

$$T = \frac{1}{2}\sum_{i=1}^{3N} m_i(\dot{x}^i)^2$$

$$\frac{\partial T}{\partial \dot{q}^i} = \sum_{j=1}^{3N} m_j \dot{x}^j \frac{\partial \dot{x}^j}{\partial \dot{q}^i} \tag{2.2.9}$$

and, using 2.2.8

$$\dot{x}^j = \frac{dx^j}{dt} = \sum_{i=1}^{3N} \frac{\partial x^j}{\partial q^i} \dot{q}^i + \frac{\partial x^j}{\partial t} \qquad (2.2.10)$$

so the required derivative on the right of 2.2.9 is

$$\frac{\partial \dot{x}^j}{\partial \dot{q}^i} = \frac{\partial x^j}{\partial q^i} \qquad (2.2.11)$$

and 2.2.9 becomes

$$\frac{\partial T}{\partial \dot{q}^i} = \sum_{j=1}^{3N} m_j \dot{x}^j \frac{\partial x^j}{\partial q^i} \qquad (2.2.12)$$

and so

$$\frac{d}{dt}\left(\frac{\partial T}{\partial \dot{q}^i}\right) = \sum_{j=1}^{3N} m_j \frac{d}{dt}\left(\dot{x}^j \frac{\partial x^j}{\partial q^i}\right) \qquad (2.2.13)$$

The right hand side of this expression can be expanded out by performing the explicit differentiation and it has been left in this "mixed" form - depending on both the q^i and the x^j - in order that certain terms may be identified and simplified by the use of Newton's Law in Cartesians. Thus

$$\frac{d}{dt}\frac{\partial T}{\partial \dot{q}^i} = \sum_{j=1}^{3N}\left(m_j \ddot{x}^j \frac{\partial x^j}{\partial q^i} + m_j \dot{x}^j \frac{d}{dt}\frac{\partial x^j}{\partial q^i}\right)$$

Now

$$m_j \ddot{x}^j = F_j \quad \text{(Newton, Cartesians)}$$

and the second term is recognisable(!) as

$$\frac{\partial}{\partial q^i} \frac{1}{2} m_j (\dot{x}^j)^2 = \frac{\partial T}{\partial q^i} \qquad (2.2.14)$$

the partial derivative of the Cartesian expression for the kinetic energy with respect to the new generalised co-ordinates. Thus 2.2.13 now becomes

$$\frac{d}{dt}\frac{\partial T}{\partial \dot{q}^i} = \sum F_j \frac{\partial x^i}{\partial q^j} + \frac{\partial T}{\partial q^i} \qquad (2.2.15)$$

And, since

$$F_j = -\frac{\partial V}{\partial x^j} \qquad (2.2.16)$$

$$\frac{d}{dt}\frac{\partial T}{\partial \dot{q}^i} = -\sum_{j=1}^{3N} \frac{\partial V}{\partial x^j} + \frac{\partial T}{\partial q^i} \qquad (2.2.17)$$

If V does not depend on t (a conservative system, as we assumed) then
the sum on the right hand side is just the transformation familiar from
the chain rule and is the partial derivative of V with respect to q^i . We
therefore have

$$\frac{d}{dt}\frac{\partial T}{\partial \dot{q}^i} = -\frac{\partial V}{\partial q^i} + \frac{\partial T}{\partial q^i}$$

or

$$\frac{d}{dt}\frac{\partial T}{\partial \dot{q}^i} = \frac{\partial}{\partial q^i}(T - V) \qquad (2.2.18)$$

Which, while reminiscent of 2.2.6 , contains an extra term: the derivative
of the kinetic energy with respect to the generalised co-ordinates.

This completes the derivation of Lagrange's equations in general co-
ordinates. There are 3N equations 2.2.18 which contain only derivatives
of the two scalar energy functions T and V. This is the form in which
Lagrange's equations are usually used to solve practical problems in particle
mechanics. The additional term in 2.2.18 $\partial T/\partial q^i$ is often referred to as a
"fictitious force"since it is responsible for the appearance of co-ordinate-
system-dependent forces like centrifugal forces, Coriolis forces etc.

From the point of view of the unification of the equations of mechanics,
however, we may use the fact that we are dealing with a conservative system
of forces one more time. If there is to be no dissipation of energy then V
cannot be a function of the velocities $\dot{q}^i$ and so

$$\frac{\partial V}{\partial \dot{q}^i} = 0 \quad (i = 1, 2, \ldots, 3N) \qquad (2.2.19)$$

so writing

$$L(q^i, \dot{q}^i) = T(\dot{q}^i, q^i) - V(q^i) \qquad (2.2.20)$$

and substituting 2.2.20 into 2.2.18 using 2.2.19 we have

$$\frac{d}{dt}\frac{\partial L}{\partial \dot{q}^i} = \frac{\partial L}{\partial q^i} \quad (i = 1, 2, \ldots, 3N) \qquad (2.2.21)$$

There is absolutely no "reason"to do this, there is no practical advantage
to be gained since in order to set up equations 2.2.21 T and V must be
known separately. It is, perhaps, worth noting at this point that, in spite
of superficial appearances to the contrary, 2.2.21 is not a partial differential equation for L but the generator of a set of 3N ordinary differential
equations for the q^i and hence $\dot{q}^i$. To solve 2.2.21 $L = T - V$ must be
known. The difference between a partial differential equation and an ordinary differential equation is crucial to certain physical interpretations of
the symbols appearing in classical and quantum mechanics: it is therefore
worth stressing that the *solution* of Lagranges equation (typical of classical

mechanical equations) is a set 3N functions $q^i(t)$ which (together with the "initial" conditions) give the explicit paths of the N particles in $3D$ space. In more precise notation the solutions are q^i

$$q^i : R^1 \to C$$

where R^1 is the real number system modelling "time" and C is a subset of R^3 which models E^3 ("ordinary space"). Further, by suitable choice of coordinate functions on E^3, C may be "unwound" so that it may be modelled by R^1; in short, C models a curve in ordinary space. In quantum mechanics the q^i are something quite different as we shall see.

These solutions of the Lagrange equation $q^i(t)$ enable us to obtain the $\dot{q}^i(t)$ at once (if the initial conditions are known) and so the formal role of the $\dot{q}^i$ is a little ambiguous in Lagrange's theory. As we have developed the theory, L has been expressed as a function of the q^i and the $\dot{q}^i$ but we have treated the $\dot{q}^i$ in the solutions as not really independent of the q^i. Fortunately, there is no ambiguity in the *mathematics* since the rules of partial differentiation used in the derivation apply whether or not the variables are independent. The usual position is to say that the q^i are the essentially independent variables and to emphasize this by using a fictional $3N$-dimensional space to model the motion of the N particles in $3D$ space called Configuration Space.

Of course, when both q^i and $\dot{q}^i$ are known as explicit functions of t we may eliminate all reference to them in the Lagrangian function by substitution:

$$L(q^i, \dot{q}^i, t) = L(q^i(t), \dot{q}^i(t), t) = \ell(t) \text{ (say)}$$

The function $\ell(t)$ is, of course, not the Lagrangian function because:

1. The domain of ℓ is "time" ie

$$\ell : R^1 \to R^1$$

   while the domain of L is the q^i and the $\dot{q}^i$

$$L : (R^1 \times R^1 \times \ldots) \times (R^1 \times R^1 \times \ldots) \to R^1$$

   and, much more important,

2. $\ell(t)$ is only defined for those q^i and $\dot{q}^i$ *which solve* 2.2.21. Although the numerical values of ℓ and L are the same when q^i and $\dot{q}^i$ solve 2.2.21 they are different functions.

The particular q^i and $\dot{q}^i$ which solve 2.2.21 are those for which the Lagrangian takes particularly important values but the Lagrangian *function*

is a function of *all possible* q^i and $\dot{q}^i$; the restriction of its domain by the
solution of equation 2.2.21 corresponds to finding those particular q^i and
$\dot{q}^i$ which are related (for a given potential function V) by Newton's Law.

These considerations enable us to set up a variation principle from which
Lagrange's equations may be derived. It is shown in Appendix 1 that, of
all the possible q^i and $\dot{q}^i$ in the domain of L, the ones which make the time
integral of L a minimum (an extremum in general) are the ones which solve
Lagrange's equations. That is, if the q^i and $\dot{q}^i$ in $L(q^i, \dot{q}^i, t)$ are allowed to
take all possible functional forms (as functions of t) then those functions
for which

$$\int L dt = \quad \text{minimum} \qquad (2.2.22)$$

are the ones which solve Lagrange's equations (and hence Newton's equations) and are therefore the ones which actually occur. Clearly, in order
to formulate such a principle at all we must be able to distinguish between
the general domain of L and the restricted domain of L which solves 2.2.21.
Equation 2.2.22 is usually written

$$\delta \int_{t_1}^{t_2} L dt = 0 \qquad (2.2.23)$$

where δ means variations in the integral induced by variations in the functional form of the q^i in the domain of L. All these points will assume
a physical and not just mathematical significance when dealing with the
quantum mechanics of particles.

We can now return to a comparison of the form of equation 2.2.21 and
its prototype in Cartesian co-ordinates 2.2.6:

$$\frac{d}{dt}\frac{\partial T}{\partial \dot{x}^i} = -\frac{\partial V}{\partial x^i} \quad (2.2.6)$$

$$\frac{d}{dt}\frac{\partial L}{\partial \dot{q}^i} = \frac{\partial L}{\partial q^i} \quad (i = 1, 2, \ldots, 3N) \quad (2.2.21)$$

The physical content of both these equations is the same: Newton's Law.
The right-hand-side of 2.2.6 is the force component and of 2.2.21 the generalised force component. The left-hand-side of both is a total time derivative
of a mechanical quantity. Clearly if the right-hand-side of either 2.2.6 or
2.2.21 is zero then this time derivative is zero and the mechanical quantity
is constant in time: it is a constant of the motion. In the intuitively-
accessible Cartesian form, the vanishing of a force component means the
vanishing of a time derivative of the kinetic energy and so the conservation
of a momentum component. Formally, equation 2.2.21 is almost identical,

the vanishing of a space-derivative of the Lagrangian function means the time-conservation of a derivative of the Lagrangian with respect to velocity. But the latter result has, as yet, no physical interpretation. What both 2.2.6 and 2.2.21 mean, at least, is that there are constants of the motion associated with those co-ordinates which do not appear in the Lagrangian function.

But the generalised co-ordinates q^i are at our disposal, we only require that they are a complete set in the sense that the Cartesian co-ordinates can be recovered from them: the Jacobian of $\partial x^j / \partial q^i$ be non-vanishing. Clearly then, either by trial and error or, preferably, by a systematic procedure it should be possible to determine the constants of the motion for particular systems from the Lagrange equations by a suitable choice of co-ordinate system. In practice, this is best done from the invariances of the variational integral 2.2.22 but in particular cases it is obvious by inspection which co-ordinates do not appear in L. This problem will be addressed later.

What remains, however, is the physical interpretation of or, at least, nomenclature for the mechanical quantities which, under appropriate conditions, become constants of the motion. In an inspired choice Hamilton choose to regard the

$$\frac{\partial L}{\partial q^i}$$

as "generalised momenta". Thus the Cartesian momenta are $\partial T / \partial \dot{x}^i$ and the generalised forces are $\partial L / \partial q^i$. Hamilton, thinking, no doubt, of the "fictitious forces" $\partial T / \partial q^i$ introduced matching "fictitious momenta"in his definition

$$p_i = \frac{\partial L}{\partial \dot{q}^i} \qquad (2.2.24)$$

This definition of momentum does two things:

1. Enables a resolution to be made of the partial ambiguity in the role of the $\dot{q}^i$ in Lagrangian mechanics

2. Paves the way for an even more general idea of how co-ordinates and momenta may be used in mechanics

But the use of 2.2.24 and its consequences (1) and (2) take us beyond Lagrangian mechanics into a more generalised approach due to Hamilton.

Summary

The basic formulation of Newtonian mechanics in Cartesian form provides the familiar intuitive concepts of "force"and "momentum". The expression

of Lagrange's form of Newton's law in general co-ordinates and its analogy
with the Cartesian form provides more generalised concepts of "force"and
"momentum". The analogy between conserved (Cartesian) momenta and
conserved generalised momenta suggests that the generalised momenta will
play a prominent role in mechanics.

2.3   Hamilton's Equations

In deciding to define the generalised momenta as

$$p_i = \frac{\partial L}{\partial \dot{q}^i}$$

and working through consequences of this definition Hamilton was making
a creative breakthrough in understanding the structure of classical mechanics. In almost all of the development which follows there is no "need"for the
generalisations of Hamilton and Jacobi in the sense that Lagrange's equations are quite sufficient for the solution of practical problems in classical
mechanics. These extensions are not done for practical reasons: the moti-
vations and justifications for them are not to be rationalised away for they
are analogy, intuition and curiosity. The mathematics of the development
is, of course, correct and easy to follow; what is not easy to appreciate is the
creative and intuitive "reasoning"which led to its development. We may ra-
tionalize the development *a posteriori* by saying that the driving force was
to greater generality in what can be considered to be a "co-ordinate"or a
"momentum". But all creations make themselves obvious in time; we can
only study the results as examples of the best work of the best minds.

The Lagrangian function L, as we have seen is (in general) a function
of the q^i , $\dot{q}^i$ and t:

$$L = L(q^i, \dot{q}^i, t) \tag{2.3.1}$$

so that

$$\frac{dL}{dt} = \frac{\partial L}{\partial t} + \sum_i \left(\frac{\partial L}{\partial q^i} \dot{q}^i + \frac{\partial L}{\partial \dot{q}^i} \ddot{q}^i \right) \tag{2.3.2}$$

This, of course, is an identity. If we now see what happens when the q^i and
$\dot{q}^i$ which satisfy Lagranges Equations are inserted in (2) i.e. those q^i and
$\dot{q}^i$ for which

$$\frac{\partial L}{\partial q^i} = \frac{d}{dt} \frac{\partial L}{\partial \dot{q}^i} \tag{2.3.3}$$

we have

$$\frac{dL}{dt} = \frac{\partial L}{\partial t} + \sum_{i=1}^{3N} [\dot{q}^i \frac{d}{dt} \frac{\partial L}{\partial \dot{q}^i} + \ddot{q}^i \frac{\partial L}{\partial \dot{q}^i}]$$

i.e.

$$\frac{dL}{dt} = \frac{d}{dt} \sum_{i=1}^{3N} \dot{q}^i \frac{\partial L}{\partial \dot{q}^i} + \frac{\partial L}{\partial t}$$

which, using the definition of the generalised momenta

$$p_i = \frac{\partial L}{\partial \dot{q}^i}$$

becomes

$$\frac{d}{dt} \left(\sum_{i=1}^{3N} p_i \dot{q}^i - L \right) = -\frac{\partial L}{\partial t} \qquad (2.3.4)$$

And, in the conservative cases which we have considered with no explicit time dependence of L on t

$$\frac{d}{dt} \left(\sum_{i=1}^{3N} p_i \dot{q}^i - L \right) = 0$$

or

$$\sum_{i=1}^{3N} p_i \dot{q}^i - L = \text{constant}$$

For those q^i , $\dot{q}^i$ (and the derived p_i) which solve the Lagrange equations.

Thus, for conservative systems, the function 2.3 is a constant of the motion and so is of general interest for all values of q^i , $\dot{q}^i$ even those which do not solve the equations of motion. In fact it is obvious that the general function given by 2.3 must be capable of being used to express the equation of motion since it contains the Lagrangian function L. Hamilton's insight was to use the function

$$\sum_{i=1}^{3N} p_i \dot{q}^i - L(q^i, \dot{q}^i, t) \qquad (2.3.5)$$

in a new way by considering it to be a function, not of the q^i $\dot{q}^i$ and the dependent p_i , but of the new variables q^i and p_i now considered to be independent. To do this we must return to the definition of the p_i :

$$p_i = \frac{\partial L}{\partial \dot{q}^i} \qquad (2.3.6)$$

that is, the p_i are functions of the q^j and the $\dot{q}^j$ (and t in general)

$$p_i = p_i(q^j, \dot{q}^j, t) \qquad (2.3.7)$$

Intuitively one feels that this relationship should be invertible (as it is trivially in the Cartesian case $p = m\dot{x}$), that is there should be a relationship of the type

$$\dot{q}^i = \dot{q}^i(q^j, p_j, t) \qquad (2.3.8)$$

re-generating the $\dot{q}^j$ from the p_i defined by 2.3.6. This is clearly possible if the Jacobian of the transformation 2.3.6 is non-zero; i.e. the determinant of

$$\frac{\partial^2 L}{\partial \dot{q}^i \partial \dot{q}^j}$$

is non-zero.

It is clear that our notation is inadequate for the task of clarifying the relationships among the various dependent and independent variables; in particular the simple notation like 2.3.7 and 2.3.8 where the same symbol is used for the quantity and the function which generates its value from other quantities is inadequate. We therefore, temporarily at least, introduce a more explicit notation and write

$$p_i = \alpha_i(q^j, \dot{q}^j, t) \qquad (2.3.9)$$

$$\dot{q}^i = \beta^i(q^j, p_j, t) \qquad (2.3.10)$$

where, for example, p_i is the ith momentum component of the system of particles and the function α_i is the way in which this momentum is generated from the co-ordinates and velocities of the particles: in fact 2.3.6.

With this notation we can now give an unambiguous formulation of Hamilton's new approach to mechanics. Hamilton used the function (6) considered as a function of the co-ordinates and momenta of the particles (and time) as the basis of his expression of the laws of mechanics i.e.

$$H(q^i, p_i, t) = \sum_{i=1}^{3N} p_i \beta^i(q^j, p_j, t) - L(q^k, \beta^k(q^j, p_j, t), t)) \qquad (2.3.11)$$

or in a more compact notation

$$H = \sum_{i=1}^{3N} p_i \dot{q}^i - L \qquad (2.3.12)$$

The function H is now universally called the Hamilton function.

The close relationship between the Lagrangian L and the Hamiltonian H means that certain identities are satisfied by partial derivatives of the two functions with respect to the independent variables relevant to each function. Retaining the full notation 2.3.11 for the Hamiltonian we have,

$$\frac{\partial H}{\partial q^i} = -\frac{\partial L}{\partial \dot{q}^j} - \sum_{j=1}^{3N} \frac{\partial L}{\partial \dot{q}^j} \frac{\partial \beta^j}{\partial q^i} + \sum_{j=1}^{3N} p_j \frac{\partial \beta^j}{\partial q^i}$$

and, using the definition of the p_j 2.3.6 we have

$$\frac{\partial H}{\partial q^i} = -\frac{\partial L}{\partial q^i} \qquad (2.3.13)$$

where, of course the left-hand-side assumes constant $q^j\,(j \neq i)$ and p_i while the right-hand-side assumes constant $q^j\,(j \neq i)$ and $\dot{q}^j$ Similarly

$$\frac{\partial H}{\partial t} = -\frac{\partial L}{\partial t} - \sum_{j=1}^{3N} \frac{\partial L}{\partial \dot{q}^j} \frac{\partial \beta^j}{\partial t} + \sum_{j=1}^{3N} p_j \frac{\partial \beta^j}{\partial t}$$

generates, by use of 2.3.6

$$\frac{\partial H}{\partial t} = -\frac{\partial L}{\partial t} \qquad (2.3.14)$$

where the right-hand-side has constant $q^i\ p_i$ and the left constant $q^i\ \dot{q}^i$. Finally,

$$\frac{\partial H}{\partial p_i} = -\sum_{j=1}^{3N} \frac{\partial L}{\partial \dot{q}^j} \frac{\partial \beta^j}{\partial p_j} + \sum_{j=1}^{3N} p_j \frac{\partial \beta^j}{\partial p_i} + \beta^i$$

gives

$$\frac{\partial H}{\partial p_i} = \beta^i = \text{``}\dot{q}^i\text{''} \qquad (2.3.15)$$

where the functional form of the velocity component in terms of co-ordinates and momenta has been replaced by the symbol for its value for graphic purposes.

These three relationships 2.3.12, 2.3.13 and 2.3.14 are *identities*, valid for all values of the arguments q^i, p_i, and t of the Hamiltonian function not, for example, just those $q^i\ p_i\ \dot{q}^i$ which solve Lagrange's equation; there is no law of mechanics amongst them since they simply flow from the mathematics of the definition of the function H.

However, Lagranges equations involve $\partial L/\partial q^i$ and 2.3.12 relates this quantity to a derivative of H. Lagranges equation can be written, therefore,

$$\frac{d}{dt}\frac{\partial L}{\partial \dot{q}^i} = \frac{\partial L}{\partial q^i} = -\frac{\partial H}{\partial q^i}$$

and, using 2.3.6

$$\frac{d}{dt}(p_i) = -\frac{\partial H}{\partial q^i}$$

i.e.

$$\dot{p}_i = -\frac{\partial H}{\partial q^i} \qquad (2.3.16)$$

This equation is identical in content to Lagranges equation and is expressed entirely in terms of functions of q^i p_i: it is Hamilton's equation of motion. Equation 2.3.15 in its "colloquial form"

$$\frac{\partial H}{\partial p_i} = \dot{q}^i \qquad (2.3.17)$$

has a similar, rather symmetrical look to 2.3.16 but it is quite different in content: 2.3.16 is a law of motion, 2.3.17 is an identity, a definition of $\dot{q}^i$ (velocity) in terms of q^i and p_i . In fact 2.3.17 "defines"velocity in terms of q^i and p_i in an analogous way to 2.3.6 being the definition of momentum in terms of q^i and $\dot{q}^i$. The two equations 2.3.16 and 2.3.17 have come to be known as Hamilton's canonical equations of motion, notwithstanding their different ontological status and the mixture of notation. Using the same notation, we can write Lagrange's equations as

$$\dot{p}_i = \frac{\partial L}{\partial q^i} \qquad (2.3.18)$$

and

$$p_i = \frac{\partial L}{\partial \dot{q}^i} \qquad (2.3.19)$$

as the ancillary definition of momentum.

But written in this way for comparison with 2.3.16 and 2.3.17 we can see the genius of Hamilton's transformation. The Law of motion 2.3.18 involves p_i q^i *and* $\dot{q}^i$ and, as we saw in the last section, the theory of Lagrange is rather ambiguous about the status of the $\dot{q}^i$ as independent variables and 2.3.18 and 2.3.19 introduce a further set of variables p_i of similar status. Compared with this Hamilton's equation 2.3.16 is, when considered in terms of independent variables, "self-contained"in that it is expressed entirely in terms of the q^i and p_i . The definitions 2.3.17 are not needed in the formulation of 2.3.16 unlike 2.3.18 and 2.3.19.

This distinction between the status of equations 2.3.16 and 2.3.17 has been emphasized because, and precisely because, further developments of mechanics are to use the freedom that 2.3.17 is an identity, independent of any laws of mechanics, to generate more generalised methods for dealing with mechanical problems. The formal analogy between 2.3.16 and

2.3.17 is too tempting to be ignored. In Hamilton's theory the independent variables are unambiguously the co-ordinates and (generalised) momentum components: $6N$ altogether. We may press the "structural argument" a little further by evaluating the total time derivative of H:

$$\dot{H} = \frac{dH}{dt} = \frac{\partial H}{\partial t} + \sum_{i=1}^{3N} \frac{\partial H}{\partial q^i} \dot{q}^i + \sum_{i=1}^{3N} \frac{\partial H}{\partial p_i} \dot{p}_i \qquad (2.3.20)$$

and substituting 2.3.16 and 2.3.17 into 2.3.20 we have, for the q^i and p_i which solve Hamiltons equations,

$$\dot{H} = \frac{\partial H}{\partial t} \qquad (2.3.21)$$

a relationship showing a remarkable formal similarity to 2.3.16. If we wish to consider p_i as "conjugate" to a corresponding q^i via 2.3.18 then we may regard H as a generalised momentum component conjugate to $-t$. This result is consistent with the relationship 2.3.17 and 2.3.14 in terms of the Lagrangian function. However, unlike 2.3.17 which is a definition, we have used 2.3.16 to obtain 2.3.21 i.e. the *law of motion* is involved in 2.3.21. This relationship between time and the Hamiltonian whose arguments solve the equations of motion was our starting point when we showed that H was conserved when the Lagrangian was independent of time. In fact it is easy to show that, provided the kinetic energy is quadratic in the velocities,

$$\sum_{i=1}^{3N} p_i \dot{q}^i = 2T$$

and so $H = E$ the total energy of the system when q^i and p_i solve 2.3.16 and 2.3.17. That is to say, since the "time - Hamiltonian" conjugacy only holds for the solutions of the mechanical problem, and the Hamiltonian is equal to the energy in this case also, it is not clear whether $-t$ is conjugate to H or to E. We will return to this problem in due course.

It is, perhaps, useful to illustrate the difference between the two types of canonical equation 2.3.16 and 2.3.17 by referring to an example in which H is particularly symmetrical in q and p. The Lagrangian for a one-dimensional harmonic oscillator is

$$L(q, \dot{q}) = \frac{1}{2} m\dot{q}^2 - \frac{1}{2} kq^2$$

The momentum p is

$$p = \frac{\partial L}{\partial \dot{q}} = m\dot{q}$$

and so the Hamiltonian is

$$H = \frac{1}{2m}p^2 + \frac{1}{2}kq^2 \ \ (= ap^2 + bq^2 \ \ say)$$

Now applying 2.3.17 we have

$$\frac{\partial H}{\partial p} = \frac{p}{m} = \dot{q}; \ \ p = m\dot{q}$$

which is not news — it is the definition of p above. But applying 2.3.16 we get

$$-\frac{\partial H}{\partial q} = kq = -\dot{p}; \ \ \dot{p} = -kq$$

which *is* news, it is Newton's Law of motion for the harmonic oscillator. However we cannot integrate

$$\dot{p} = -kq$$

and solve the mechanical problem since p and q are independent variables: we cannot express p as a function of q to complete the solution. This is where the result of applying 2.3.17 comes in: $p = m\dot{q}$ from 2.3.17 provides the relationship between p and q which enables the law of motion to be integrated:

$$\dot{p} = m\ddot{q} = -kq$$

can now be solved by direct quadrature. To stress the point again, 2.3.16 is the law of motion which incorporates Newton's law and provides those p's and q's which solve the mechanical problem; while 2.3.17 is an identity, true for all p's and q's which establishes the connection between the p's and q's enabling the law of motion equation to be integrated. Of course, in order to obtain a complete solution of the canonical equations for a particular case one must have a set of "initial conditions": initial positions and momenta of the particles, a set of $2n$ constants where n is the number of pairs of variables p and q. For a system of N particles in ordinary space $2n = 6N$.

The reason for the appearance of two canonical equations rather than one is precisely because the q^i and p_i are considered to be independent; leading to twice as many degrees of freedom in the problem and so twice as many equations are needed to fix the motions of the particles. This point is best brought out in the variational formulation of Lagrange's and Hamiltons equations which is sketched in the Appendix; the essential point is that in Lagrange's equation the $\dot{q}^i$ are treated as derivatives of the q^i and this relationship is treated explicitly by an integration whereas in Hamilton's formulation, no such dependence of the p_i on the q^i is assumed.

It is becoming increasingly obvious that the analytic treatment of mechanics admits a "geometric"analogy. We may consider the variables of the mechanics of a system of N particles as a many-dimensional "manifold". These are $3N$ q^i, $3N$ p_i, H and t : $6N + 2$ altogether. The identities 2.3.17 and 2.3.21 provide $3N + 1$ constraints and the equation of motion 2.3.16 provides another $3N$. We therefore have $6N + 1$ equations to be satisfied in our $6N + 2$ dimensional "space"generating a unique "curve"in this fictional space. This analogy means that many of the powerful *techniques* of differential geometry may be used in the mathematical treatment of classical particle mechanics. But there is no physics in this as the $6N + 2$ dimensional space does not exist, what is really happening is that N particles are moving in N paths in real (existing) $3D$ space and the one-one mapping between these paths and a single curve in a $6N + 2$ dimensional space is a mathematical construct.

Finally, as we insisted of the Lagrangian function L, it must be stressed that the domain of the Hamilton function is the whole of the $6N$-dimensional space spanned by the q^i and the p_i (plus possibly t), not just the space spanned by those q^i and p_i which solve the canonical equations of motion; for, when those $6N$ equations are solved, the Hamiltonian is fixed. If the Hamiltonian has no explicit time dependence it is fixed as a constant (the total energy); if the Hamiltonian has explicit time dependence it is fixed as a *function of t only* by the constraints of the canonical equations. In short, the Hamiltonian is a function of the q^i and p_i and the equations of motion are constraints on the values which it may take if H is to be evaluated only at points on the real paths of the particles. This important point admits the possibility of variational principles which consider q^i and p_i which do not solve the equations of motion, but it has a much more important physical role to play in the formulation of the quantum mechanics of particles.

Summary

One's feelings on meeting Hamilton's formulation of classical particle mechanics continued in the canonical equations are usually ones of suspicion and puzzlement at the way he was able to double the number of independent variables in the problem with apparent impunity. While Lagrange's equations contain q^i and $\dot{q}^i$, ultimately one is in no doubt that the $\dot{q}^i$ are dependent on the q^i; even the notation suggests it. In deriving Lagrange's equations from a variation principle the $\dot{q}^i$ are expressed in terms of the q^i. These feelings are reinforced by any familiarity with the use of Hamilton's equations to solve practical problems : the first step is always, as we have seen in the simple harmonic oscillator case, to *eliminate* half of the variables (the p's) and proceed with equations which are then substantially the same as Lagrange's equations.

The *coup de grace* occurs when the canonical equations are derived from a variation principle in which independent variations in the q^i and p_i are allowed. The variation principle generates *both* sets of canonical equations

$$\dot{p}_i = -\frac{\partial H}{\partial q^i} \quad : \quad \dot{q}^i = \frac{\partial H}{\partial p_i}$$

but the second set can be derived by elementary manipulations from the definition of H in terms of L , p_i and $\dot{q}^i$: they are, in fact, identities *independent* of the variation principle which generates them. What has happened is that, in the particular case of H having the dependence it does on the variables q^i p_i , the q^i and p_i can be treated as independent (when they are not) precisely because the variation principle requires a set of equations to be satisfied which are, in reality, satisfied identically by the particular H.

That is, if the particles satisfied some dynamical law other than Newton's Law and so "H" would have a different form. In this case the identity satisfied between $\dot{q}^i$ and "H" might well be different from (and therefore incompatible with) the substantive connection between q^i and "H" required by a variation principle. In this hypothetical case it would not be possible to treat the q^i and p_i as independent without contradiction. So Hamilton's formulation with q^i and p_i treated as independent is really "sailing with the current" of Newtonian mechanics.

2.4 Transformation Theory

The solution of Hamilton's canonical equations

$$\frac{\partial H}{\partial q^i} = -\dot{p}_i \tag{2.4.1}$$

$$\frac{\partial H}{\partial p_i} = \dot{q}^i \tag{2.4.2}$$

is obviously enormously simplified if the Hamiltonian function is *independent* of some of the co-ordinates q^i; since in this case the left-hand-side of 2.4.1 is zero and so $\dot{p}_i = 0$ (p_i=constant) integration is immediate. Similar comments hold for the identities 2.4.2. Clearly, if a co-ordinate system could be found for which the Hamiltonian was independent of all co-ordinates then the solution of the equations of motion would be trivial: all momenta would be constants. Before spending too much time in searching for such a co-ordinate system - which would obviously depend on the potential in which the particles move - consider the physical interpretation of such a system.

If all momenta are constants then, because of Newton's law there are no forces acting on the particles (or bodies in general) and so, apparently, this co-ordinate system can only be found for the trivial case of non-interacting particles in free motion (or free rotation of extended bodies). The only way in which one could hope to generate a co-ordinate system with the sort of properties we require for non-trivial mechanical problems is to allow the origins of the co-ordinates to move and "follow"the particles; so that the momenta in the moving co-ordinate system can be constant in the presence of a potential and inter-particle interaction. Now, in the Hamiltonian formulation of mechanics we have an ideal vehicle for the construction of such moving co-ordinate systems; the co-ordinate and momenta are independent variables and the canonical equations 2.4.1 and 2.4.2 are formally almost identical. So, if we admit transformation of co-ordinates and momenta which "mix"the original co-ordinates and momenta in the definition of the new sets we should be able to generate a set of canonical equations in terms of the new variables which do have the desirable properties. That is, we seek a transformation to new variables Q^j, P_j such that

$$Q^j = Q^j(q^i, p_i, t) \tag{2.4.3}$$
$$P_j = P_j(q^i, p_i, t) \tag{2.4.4}$$

for which

$$\frac{\partial h}{\partial Q^j} = -\dot{P}_j = \text{ constant} \tag{2.4.5}$$

$$\frac{\partial h}{\partial P_j} = \dot{Q}^j \tag{2.4.6}$$

where h is the Hamiltonian in terms of the variables Q^j , P_j .

Now, there are several points to consider when setting out on such a venture:

1. The transformations 2.4.3 should be a complete and non-redundant set if the original q^i and p_i were; this condition can be expressed in the usual way as the non-vanishing of a Jacobian.

2. The transformation 2.4.3 must be such that the transformed equations of motion stay within the canonical Hamiltonian formalism i.e. are indeed of the type 2.4.5 and 2.4.6.

3. Since the transformation 2.4.3 "mixes"co-ordinates and momenta of the original "intuitive"type, the new, transformed, equations 2.4.5

and 2.4.6 cannot be so easily distinguished as the original equations
2.4.1 and 2.4.2. That is, while 2.4.1 is an equation of motion and
2.4.2 is an identity in the original formulation, in the transformed
equations the "equations of motion"may be inextricably intermingled
with the "identities"and so we are forced to treat equations 2.4.3 and
2.4.5 and, by implication, 2.4.1 and 2.4.2 on the same footing. If we
do this then we may as well be hung for a sheep as for a lamb and
seek transformations which make both P_i and Q^i constants.

It must be said at this point that the equations describing the motions of
a system of particles cannot be solved by sleight of hand. In attempting
to find a particularly trivial form of the equations of motion by use of a
transformation we are simply pushing the difficulties "out of sight"into the
generation of the transformation. None of these manipulations are being
carried through as aids to the practical solution of problems in mechanics;
our aim is to obtain the most general form of the mechanical principles in
order to throw light on the formalism and physical interpretation of the
quantum mechanics of systems of particles.

There are no intuitive guides to be used in seeking the transformations (3) and so we must fall back on general principles. The most general
formulation of Hamiltonian equations is the co-ordinate-free variational formulation sketched in the Appendix:

$$\delta \int_{t_1}^{t_2} [\sum_{i=1}^{3N} p_i \dot{q}^i - H]dt = 0 \qquad (2.4.7)$$

from which the canonical equations 2.4.1 and 2.4.2 follow. If we wish,
therefore, to use only transformations 2.4.3 which remain in the Hamiltonian formalism we require

$$\delta \int_{t_1}^{t_2} [\sum_{i=1}^{3N} P_i \dot{Q}^i - h]dt = 0 \qquad (2.4.8)$$

in addition to 2.4.7. Now in Appendix 1 the solution of problems like 2.4.7
were considered *for cases where the integrand is a known function* of the
arguments and we sought the optimum functional form of those arguments
as functions of t. But the variational problem

$$\delta \int_{t_1}^{t_2} f dt = 0 \qquad (2.4.9)$$

can always be solved by a function whose total time derivative is f, i.e. if

$$\frac{dF}{dt} = f$$

then

$$\delta \int_{t_1}^{t_2} f dt = \delta(F(t_2) - F(t_1)) = 0$$

identically. Or, what amounts to the same thing, the variation principle only determines the optimising integrand to within an additive total time derivative. This means, of course, that when f is a *known function* of some arguments other than just t, like q^i, q^i, p_i in

$$\sum_{i=1}^{3N} p_i \dot{q}^i - H(q^j, p_j)$$

then, when the variational problem is solved, the integral which solves the problem is a function of t only. That is, of course, that (in our case) q^i and p_i are given as known functions of t which, if we choose to do it, we may substitute in the integrand for q^i and p_i and obtain

$$\frac{dG(t)}{dt} = \sum_{i=1}^{3N} p_i \ddot{q}^i - H(q^j, p_j)$$

explicitly.

We can do this for both the original variables in 2.4.7 and the transformed variables in 2.4.8 and combine the two results; both of which are functions of t only so we may combine 2.4.7 and 2.4.8 to give

$$\delta \int_{t_1}^{t_2} [(\sum_{i=1}^{3N} p_i \dot{q}^i - H) - (\sum_{i=1}^{3N} P_i \dot{Q}^i - h) dt] = 0$$

It is clear that the integrand can be set equal to the total time derivative of an arbitrary function F (say)

$$[\sum_{i=1}^{3N} p_i \dot{q}^i - H] - [\sum_{i=1}^{3N} P_i \dot{Q}^i - h] = \frac{dF}{dt} \qquad (2.4.10)$$

In obtaining this equation we have put no requirements on the transformation 2.4.3, we have not yet sought a transformation which will generate constant Q^i and P_i. However, it is clear that F contains a characterisation of the transformation since it is a function of the q^i, p_i, Q^i and P_i; the question is "how do we extract 2.4.3 from 2.4.10"?

First of all, although F is a function of the q^i, p_i, Q^i, P_i (and t), only $6N$ of these $12N$ variables are independent because of 2.4.3. We may choose which $6N$ at our pleasure. To illustrate the procedure we choose q^i and Q^i

$$F = F(q^i, Q^i, t)$$

so that

$$\frac{dF}{dt} = \sum_{i=1}^{3N} \frac{\partial F}{\partial q^i} \dot{q}^i + \sum_{i=1}^{3N} \frac{\partial F}{\partial Q^i} \dot{q}^i + \frac{\partial F}{\partial t} \qquad (2.4.11)$$

Now both 2.4.10 and 2.4.11 are identities and may be combined to give

$$\sum_{i=1}^{3N} (p_i - \frac{\partial F}{\partial q^i}) \dot{q}^i + \sum_{i=1}^{3N} (P_i + \frac{\partial F}{\partial Q^i}) \dot{Q}^i + (h - H - \frac{\partial F}{\partial t}) = 0 \qquad (2.4.12)$$

which itself is an identity so that the individual coefficients of the $6N + 1$ time derivatives must be separately zero:

$$p_i = \frac{\partial F(q^j, Q^j, t)}{\partial q^i} \qquad (2.4.13)$$

$$P_i = -\frac{\partial F(q^j, Q^j, t)}{\partial Q^i} \qquad (2.4.14)$$

$$(h - H) = \frac{\partial F(q^j, Q^j, t)}{\partial t} \qquad (2.4.15)$$

Equation 2.4.13 fixes the Q^i in terms of the q^i and p_i and may be solved to generate their explicit form. Once the Q^i are found from 2.4.13 they may be substituted in 2.4.14 to generate the P_i and in 2.4.15 to obtain the new Hamiltonian h.

These manipulations show that functions like F may be used to generate transformation of the variables in the canonical equations; what is not yet clear is how to choose a particular F which simplifies the canonical equation in order to generate 2.4.5 and 2.4.6. The key lies in the back-substitution of 2.4.13, 2.4.14 and 2.4.15 into the canonical equations which generates an equation for F.

We now suppose that the transformation of co-ordinates and momenta in which the new co-ordinates and momenta Q^i, P_i are all constants can be found and express it in terms of an F of the above type. In deference to long-established practice we call this particular F, S and it is a function of the independent "variables" q^i, Q^i, t although the Q^i are to be constants ultimately, albeit "independent constants".

$$S = S(q^i, Q^i, t)$$

We require S such that the transformed Hamiltonian h satisfies

$$\frac{\partial h}{\partial P_i} = 0 : \quad \frac{\partial h}{\partial Q^i} = 0 \qquad (2.4.16)$$

If the transformed Hamiltonian h contains no explicit time dependence we have

$$\frac{\partial h}{\partial t} = 0 \qquad (2.4.17)$$

and so the Hamiltonian is a constant, having no dependence on Q^i, P_i or t. Now the original momenta p_i may be generated from S by the use of 2.4.13 so we may write

$$H(q^i, p_i) \quad \text{as} \quad H(q^i, \frac{\partial S}{\partial q^i})$$

and, substituting into 2.4.15, we have

$$H(q^i, \frac{\partial S}{\partial q^i}) + \frac{\partial S}{\partial t} = h \qquad (2.4.18)$$

where h (the transformed Hamiltonian) is a constant as we have seen above. This is a partial differential equation for S since we are now going to regard S as a function of the q^i and t only since the Q^i are constants. In fact a trivial re-definition of S enables us to absorb the constant h, replacing S by $S - Et$ enables the equation to be written in a compact form

$$H(q^i, \frac{\partial S}{\partial q^i}) + \frac{\partial S}{\partial t} = 0 \qquad (2.4.19)$$

where now, $S = S(q^i, t)$ This equation, the Hamiltonian-Jacobi equation, can be set up whenever the Hamiltonian can be formed in terms of the original q^i and p_i. If it can be solved the mechanical problem is solved.

This equation is a *partial differential equation* and its physical interpretation must be sought since it is the "high point"of classical mechanics in the sense that it is in the Hamiltonian-Jacobi equation where, historically and methodologically, classical mechanics is closest to quantum mechanics. But, before seeking a physical interpretation it is illuminating to give more direct derivation of the Hamilton-Jacobi equation which does not have to make the all-important *assumption* that the solution S actually exists: that such a specialised transformation can indeed be found.

In the case where the transformed Hamiltonian function is a constant, which as we have seen may be chosen to be zero by a slight re-definition of S, we can go back to the original equation 2.4.10 which defined the original general transformation:

$$[\sum_{i=1}^{3N} p_i \dot{q}^i - H] - [\sum_{i=1}^{3N} P_i \dot{Q}^i - h] = \frac{dF}{dt}$$

If the Q^i are constants then $\dot{q}^i = 0$ and, if h is chosen to be zero, we have for the special case $F = S$:

$$\frac{dS}{dt} = [\sum_{i=1}^{3N} p_i \dot{q}^i - H] = L$$

That is

$$S = \int L dt$$

showing the special relationship between this particular transformation function and the original variational formulation of Lagrange's equations. The variational principle is now

$$\delta S = 0$$

It should be noted that this derivation of the transformation equations in general and the Hamilton-Jacobi equation in particular has *used the dynamical law* by assuming that F (or S) is a function of time only. Thus these are indeed equations not identities.

2.5   The Hamilton-Jacobi Equation

In the variational principle used in this chapter and outlined in Appendix 1, the paths of the particles have been allowed to vary i.e. the δq^i are arbitrary functions of t. The variations in the velocities $\delta \dot{q}^i$ are then fixed as the time derivatives of these path variations. However, if we allow the time parameter to vary as well as the paths we can obtain a more general type of variation, the so-called Δ variation. We write

$$q^i \rightarrow q^i + \Delta q^i = q^i + \delta q^i + \dot{q}^i \Delta t$$

where δq^i has the same meaning as before. The relationship between Δ and δ as symbolic "operators" is a particularly simple one as we can see from the variation of a function of $q^i, \dot{q}^i$ and t: $f(q^i, \dot{q}^i, t)$.

$$\Delta f = \sum_{i=1}^{3N} \frac{\partial f}{\partial q^i} \Delta q^i + \sum_{i=1}^{3N} \frac{\partial f}{\partial \dot{q}^i} \Delta \dot{q}^i + \frac{\partial f}{\partial t} \Delta t$$

$$= \sum_{i=1}^{3N} \frac{\partial f}{\partial q^i} (\delta q^i + \dot{q}^i \Delta t)$$

$$+ \sum_{i=1}^{3N} \frac{\partial f}{\partial \dot{q}^i} (\delta \dot{q}^i + \ddot{q}^i \Delta t) + \frac{\partial f}{\partial t} \Delta t$$

$$= \sum_{i=1}^{3N} \frac{\partial f}{\partial q^i} \delta q^i + \sum_{i=1}^{3N} \frac{\partial f}{\partial \dot{q}^i} \delta \dot{q}^i$$

$$+ \Delta t [\sum_{i=1}^{3N} \frac{\partial f}{\partial q^i} \dot{q}^i + \sum_{i=1}^{3N} \frac{\partial f}{\partial \dot{q}^i} \ddot{q}^i + \frac{\partial f}{\partial t}]$$

$$= \delta f + \frac{df}{dt} \Delta t$$

$$i.e. \quad \Delta f = \delta f + \Delta t \frac{df}{dt}$$

or, symbolically,

$$\Delta = \delta + \Delta t \frac{d}{dt} \tag{2.5.1}$$

If this new type of variation is used on

$$S = \int_{t_1}^{t_2} L dt \tag{2.5.2}$$

we have

$$\Delta S = \delta \int_{t_1}^{t_2} L dt = \delta \int_{t_1}^{t_2} L dt + [L \Delta t]_{t_1}^{t_2} \tag{2.5.3}$$

As before, using the δ variation

$$\delta \int_{t_1}^{t_2} L dt = \int_{t_1}^{t_2} \sum_{i=1}^{3N} (\frac{\partial L}{\partial q^i} \delta q^i + \frac{\partial L}{\partial \dot{q}^i} \delta \dot{q}^i) dt \tag{2.5.4}$$

But this time this term is not zero because the time is being varied as well as the paths. To reduce this expression further we use two results. Firstly we note that the δ variation commutes with differentiation as usual

$$\frac{d}{dt} \delta f = \delta \frac{df}{dt} = \delta \dot{f}$$

i.e.

$$\delta \dot{q}^i = \frac{d}{dt} \delta q^i \tag{2.5.5}$$

a result used in the reduction of the δ variation in Appendix 1. Much more importantly, we may use the result of the fixed-time δ variation i.e. we may use Lagrange's Equations

$$\frac{\partial L}{\partial q^i} = \frac{d}{dt} \frac{\partial L}{\partial \dot{q}^i} \tag{2.5.6}$$

in the integrand of 2.5.4. This integrand then becomes (for each i)

$$\frac{\partial L}{\partial q^i}\delta q^i + \frac{\partial L}{\partial \dot{q}^i}\delta \dot{q}^i = \frac{d}{dt}\frac{\partial L}{\partial \dot{q}^i}\delta q^i + \frac{\partial L}{\partial \dot{q}^i}\frac{d}{dt}(\delta q^i)$$

$$= \frac{d}{dt}\left(\frac{\partial L}{\partial \dot{q}^i}\delta q^i\right)$$

and, since the generalised momenta are defined by

$$p_i = \frac{\partial L}{\partial \dot{q}^i}$$

the integrand becomes (for each i)

$$\frac{\partial L}{\partial q^i}\delta q^i + \frac{\partial L}{\partial \dot{q}^i} = \frac{d}{dt}(p_i\delta q^i) \tag{2.5.7}$$

Thus, for variations δ including time variations we have

$$\delta \int_{t_1}^{t_2} L\,dt = \sum_{i=1}^{3N}\int_{t_1}^{t_2}\frac{d}{dt}(p_i\delta q^i)dt = \sum_{i=1}^{3N}\left[p_i\delta q^i\right]_{t_1}^{t_2} \tag{2.5.8}$$

Giving finally,

$$\Delta S = \Delta\int_{t_1}^{t_2} L\,dt = \left[\sum_{i=1}^{3N}p_i\delta q^i\right]_{t_1}^{t_2} + [L\delta t]_{t_1}^{t_2} \tag{2.5.9}$$

for the full variation of the Lagrange integral.

It is important to note here that, in reducing 2.5.4 to 2.5.8 Lagrange's Equations of motion were used that is, while 2.5.4 is an expression involving *any* q^i , $\dot{q}^i$; using 2.5.6 means that thereafter, and in particular in 2.5.8 and 2.5.9, *only those q^i and $\dot{q}^i$ are allowed which actually solve the Lagrange equations of motion.* That is, 2.5.9 is only true for q^i and p_i along a real trajectory; q^i and p_i are assumed in 2.5.8 and 2.5.9 to be (in principle known) functions of t.

Equation 2.5.9 may be re-cast in a Hamiltonian form by use of

$$L = \sum_{i=1}^{3N}p_i\dot{q}^i - H$$

this gives

$$\Delta S = \left[\sum_{i=1}^{3N}p_i\delta q^i\right]_{t_1}^{t_2} + \left[\sum_{i=1}^{3N}p_i\dot{q}^i\Delta t - H\Delta t\right]_{t_1}^{t_2}$$

and, since

$$\Delta q^i = \delta q^i + \dot{q}^i \Delta t$$

$$\Delta S = [\sum_{i=1}^{3N} p_i \delta q^i - H \delta t]_{t_1}^{t_2} \qquad (2.5.10)$$

or

$$\Delta S = \sum_{i=1}^{3N} (p_i)_{t_2} (\Delta q^i)_{t_2} - \sum_{i=1}^{3N} (p_i)_{t_1} (\delta q^i)_{t_1} + H_{t_1} (\Delta t)_{t_1} - H_{t_2} (\Delta t)_{t_2} \quad (2.5.11)$$

Now "t_1"and "t_2"are being used simply as symbolic indications of the end points of the variation process: as limits of the integral. Now that the time integration has been carried through we may use "$t_1 = t_0$"as the initial point, and "$t_2 = t$ "as the final point giving ΔS as a function of initial (fixed) conditions and "running"conditions and enabling one set of subscripts to be dropped. The result also looks a good deal neater if the convention of summing over repeated indices is temporarily adopted:

$$\Delta S = p_i \Delta q^i - p_{i0} \Delta q_0^i + H_0 (\Delta t_0) - H \Delta t \qquad (2.5.12)$$

From this expression by partial differentiation in each of the $12N + 2$ variables $q^i, q_0^i, p_i, p_{i0}, t$ and t_0 we may obtain

$$\frac{\partial S}{\partial t} = -H \quad : \quad \frac{\partial S}{\partial q^i} = p_i \qquad (2.5.13)$$

$$\frac{\partial S}{\partial t_0} = H_0 \quad : \quad \frac{\partial S}{\partial q_0^i} = -p_{i0} \qquad (2.5.14)$$

In fact, the "initial"values t_0, q_0^i, p_{i0} are constants and the relationships 2.5.14 simply serve to indicate that, for example, of the $6N$ constants q_0^i, p_{i0} only $3N$ are independent; the form of the occurrence of the q_0^i fixes the p_{i0} . The relationships 2.5.13 may be used to generate the Hamilton-Jacobi equation:

$$H(q^i, p_i, t) = H(q^i, \frac{\partial S}{\partial q^i}, t)$$

and

$$H(q^i, \frac{\partial S}{\partial q^i}, t) = -\frac{\partial S}{\partial t}$$

or

$$H(q^i, \frac{\partial S}{\partial q^i}, t) + \frac{\partial S}{\partial t} = 0 \qquad (2.5.15)$$

which is the desired result.

This equation is a partial differential equation in $3N + 1$ variables (the q^i and t) and so, when solved, will require $3N + 1$ constants or initial conditions. One of those will be additive (corresponding to the time derivative) leaving just $3N$ required values at the initial values of the q^i : the q^i_0 . The relationships 2.5.14 may then be used to generate the initial momenta giving all $6N$ initial conditions. This result agrees with the deliberations of the last section where an identical equation was derived based on the initial assumption of the existence of a function which would transform between fixed co-ordinates (and momenta) and general co-ordinates (and momenta).

In obtaining equation 2.5.15 use was made of the dynamical Law of classical mechanics in the form of Lagrange's equations and so the final equation can only apply for trajectories which solve Lagrange's equations: i.e. for trajectories which obey Newton's Law. The same is true of the derivations from transformation theory in the last section: in this case the restriction to real allowed trajectories is less obvious in that it is implicit in the assumption that initial and final states of the transformation are both functions of time only (not of q^i or p_i), that is the variational condition is assumed to be fulfilled. Thus 2.5.15 is not an identity; it is an equation equivalent to Hamilton's canonical equations or to Lagrange's equations or indeed to $F = ma$.

But 2.5.15 is a partial differential equation for S in which the q^i are the independent variables *along with* t; there is no question in 2.5.15 that the q^i are functions of t; indeed such a thing is impossible by the very nature of the equation: it is S which is to be determined by the Hamilton-Jacobi equation not the q^i and t! The Hamilton-Jacobi equation is a *partial* differential equation unlike, for example, Hamilton's canonical equations which are (or which generate) *ordinary* differential equations. That is, unlike the canonical equations

$$\frac{\partial H}{\partial q^i} = -\dot{p_i} \quad : \quad \frac{\partial H}{\partial p_i} = \dot{q^i}$$

which involve a *known* function H (the partial derivatives merely picking out ordinary differential equations), the Hamilton-Jacobi equation involves an *unknown* function S to be determined by this equation. But this S has values for *all* q^i (and t). That is S, although it contains the dynamical law, does not determine trajectories as functions of t directly as, for any given choice of values of the q^i and t, S *has some value.*

This raises two questions; one mathematical and one scientific:

1. Precisely how does a knowledge of S determine the allowed particle trajectories?

2. What is the *referent* of S and hence of the Hamilton-Jacobi equation which determines S? To what do the solutions refer — how does one *interpret* the Hamilton-Jacobi equation?

In the past attention has been concentrated almost exclusively on the first of these points and, fortunately, the answer to (1) helps with the consideration of (2)

If, in the usual way, the initial conditions of a problem in classical particle mechanics are known this information can always be brought into the form of a knowledge of the q^i and p_i at some time t_0 : that is the q_0^i and p_{i0} of equation 2.5.14. If we solve the Hamilton-Jacobi equation we may use the q_0^i to obtain a complete solution (together with the energy at t_0 as an additive constant). Knowing S and its dependence on the q^i and q_0^i we can extract the q^i as a function of q_0^i and p_{i0} if we can invert the relations

$$p_{i0} = -\frac{\partial S(q^j, q_0^j, t, E_0)}{\partial q_0^i} \qquad (2.5.16)$$

Now, however difficult this might prove to be in practice, it is clearly possible in principle provided the Jacobian of the transformation

$$p_{i0} = f_i(q^j, q_0^j, t, E_0) \;\; : \;\; f_i = -\frac{\partial S}{\partial q_0^i} \qquad (2.5.17)$$

from the p_{i0} to the q^j is not zero. Of course the p_{i0} themselves are independent since the q_0^i are independent in order that S be a complete solution of the partial differential equation. That is, one can obtain a trajectory from S by inverting 2.5.16 provided the Jacobian of

$$\frac{\partial^2 S}{\partial q_0^j \partial q^i}$$

is not zero.

Therefore, by choosing a set of initial q_0^i and E_0 one may obtain a trajectory in configuration space; that is a set of particle trajectories in real space. So that families of trajectories of the particles may be found by suitable families of initial conditions. The key difference between the canonical equations of Hamilton and the partial differential Hamilton-Jacobi equation is that the solution of the latter contains *all possible* trajectories in configuration space (and all just once) which are distinguished by the values of the initial conditions. The trajectories appear by *inverting* a relationship involving S rather than directly as $q^i(t)$.

In discussing Lagrange's equations earlier in this chapter we noted that, in the case of a single particle, the quantities q^i appearing as solutions of these equations (or the canonical equations) were functions

$$q^i \; : \; R^1 \to C \quad C \subset R^3 \tag{2.5.18}$$

where R^1 is the real number system (modelling "time") and C a subset of R^3 (which models ordinary space) which would normally be capable of being parametrised by R^1 : C models a curve in ordinary space. Now from what we have said above it is clear that this is not what q^i is in the Hamilton-Jacobi theory. In fact, in this equation q^i is independent of t - it is a co-ordinate variable which maps points in ordinary space into the real number system: given a point in space q^i is a function which gives its numerical value in some frame of reference:

$$q^i \; : \; E^3 \to R^1 \tag{2.5.19}$$

So that, abbreviating 2.5.18 to

$$q^i \; : \; R^1 \to R^3 \quad \text{(Lagrange, Hamilton)}$$

and taking the liberty of (temporarily!) identifying $E^3 and R^3$

$$q^i \; : \; R^3 \to R^1 \quad \text{(Hamilton-Jacobi)}$$

showing that the change from the canonical equations to the Hamilton-Jacobi equation has turned q^i "inside out".

The meaning of q^i has changed from "a point on an allowed trajectory"to "a point in space"since, with the introduction of S, all points in space lie on allowed trajectories: what distinguishes between allowed trajectories is not the solution of the mechanical equations but only the initial conditions. Just as the referent of the q^i has changed so has the referent of the whole mechanical theory with the introduction of the Hamilton-Jacobi approach. The referent now is, arguably at least, *the ensemble of all possible trajectories for the given field of force and inter-particle interactions*. This provides an answer to the question (2) posed earlier about the referent of the Hamilton-Jacobi equation and its solution S. The idea of an ensemble of *one each* of all trajectories consistent with a given field of force and differing in initial conditions is a key one in the development of Schrödinger's quantum theory and is, at least incipiently, present in the high point of classical mechanics.

Of course, a series of purely mechanical manipulations with equations cannot induce the equations to change their referent and the meaning of the symbols involved to change; but the arguments and conclusions presented here can be made more acceptable using the method of "characteristic strips"in the theory of the equivalence of some partial differential equations to sets of ordinary differential equations.

2.6 Conditions on Canonical Co-ordinates

In everything which has been discussed so far the terms "co-ordinates"and "momenta"have been used rather informally insofar as we have relied on intuition and ordinary practice to supply a picture of what is meant by a co-ordinate, and the definition

$$p_i = \frac{\partial L}{\partial \dot{q}^i} \tag{2.6.1}$$

to provide a conjugate momentum component. The simplest examples show that this usage seems justified; in particular the familiar results in Cartesian co-ordinates are all consistent with this general theory. In normal practice "co-ordinate"at its most complicated usually means a member of one of the familiar 11 orthogonal co-ordinate systems in 3-dimensional space. If one of these co-ordinate systems is used, then 2.6.1 provides the conjugate momentum components which are, however, intuitively less accessible.

Quite independently of the Lagrangian and Hamiltonian formalisms there are existing definitions of various types of momenta and so it is natural to inquire about the relationship between what one might call "naturally occurring"momenta and their components and the momentum components conjugate to sets of co-ordinates. That is, under what conditions is a "co-ordinate"suitable for use in the canonical formalism and under what conditions can a "pre-existing"momentum component be made conjugate to some co-ordinate in the canonical formalism? Naturally this investigation is limited to "co-ordinates"in the original sense of the term: i.e. specifically excluding transformations which "mix"co-ordinates and momenta.

Elementary considerations are enough to show that there is a whole *class* of momentum components which cannot be brought into the canonical formalism in spite of their utility and familiarity.

In elementary (vectorial) mechanics one defines the angular momentum vector as

$$\vec{\ell} = \vec{r} \times \vec{p} \tag{2.6.2}$$

and this co-ordinate-free definition implies that the angular momentum vector can be resolved into components in any co-ordinate system whether or not any member of the co-ordinate system is an angle. That is, 2.6.2 is not concerned with the idea of momentum components conjugate to co-ordinates with the dimensions of angle (i.e. no dimensions) it is simply *called* the angular momentum for intuitively justifiable reasons. In fact, the most usual form of resolution of $\vec{\ell}$ is

$$\vec{\ell} = \ell_x \vec{i} + \ell_y \vec{j} + \ell_z \vec{k}$$

in Cartesian components and, of course ℓ_x is not conjugate to x.

What is very clear is that $\vec{\ell}$ cannot be expressed in a way in which, to each of of three linearly independent components (in some co-ordinate system), there is a conjugate angular co-ordinate. This is trivially true simply because the position of a point in space cannot be specified by three angles: at least one length is required. One might think of this simple example as suggesting a kind of "inverse problem in canonical co-ordinates": given a set of momentum components, under what conditions can are find conjugate co-ordinates which:

1. Are a complete and non-redundant set for the problem in hand,

2. Regenerate the given momentum components via 2.6.1

Now among the familiar orthogonal co-ordinate systems there are some which have two angles and one length as their dimensions. But the problem with the resolution of the angular momentum vector is more acute than we have suggested: as we shall see shortly, it is possible to make *only one* of the three components of $\vec{\ell}$ into a canonically conjugate momentum. Perhaps the most direct explanation of why this is so is by way of an explicit example: the transformation between Cartesian and spherical polar co-ordinates.

Taking

$$\vec{p} = m\vec{v} = m\dot{\vec{r}}$$

we have

$$\vec{\ell} = \vec{r} \times \vec{p} = m(\vec{r} \times \dot{\vec{r}})$$

and the Cartesian components of $\vec{\ell}$ are:

$$\ell_x = m(y\dot{z} - z\dot{y})$$
$$\ell_y = m(x\dot{z} - z\dot{x})$$
$$\ell_z = m(x\dot{y} - y\dot{x})$$

using

$$x = r \sin\theta \cos\phi$$
$$y = r \sin\theta \sin\phi$$
$$z = r \cos\theta$$

gives

$$\ell_x = -I(\dot{\theta}\sin\phi + \dot{\phi}\cos\theta\sin\theta\cos\phi)$$
$$\ell_y = I(\dot{\theta}\cos\phi - \dot{\phi}\cos\theta\sin\theta\sin\phi)$$
$$\ell_z = I_\phi\dot{\phi} = (I\sin^2\theta)\dot{\phi}$$
$$where \ \ I = mr^2$$

where I_ϕ is the "moment of inertia" of a particle about the z axis. Now for a Lagrangian of the simple form

$$L = \frac{1}{2}mv^2 - V = \frac{1}{2}m(\dot{r}^2 + r^2\dot{\theta}^2 + r^2\sin^2\theta\dot{\phi}^2) - V$$

$$\frac{\partial L}{\partial \dot{\phi}}$$

is indeed $I_\phi\dot{\phi}$ and so the angular variable ϕ and the angular momentum component

$$p_\phi = I_\phi\dot{\phi}$$

are indeed conjugate. But ℓ_z is the *z-component* of the original angular momentum not the ϕ -component!

Of course, ϕ is "tied" to the choice of z-direction (it is an angle of to 0 to 2π around the z-axis) and we could have defined ϕ in an analogous way around the x-axis and so obtained $\ell_x = I_\phi\dot{\phi}$. But we cannot do *both* if only for the simple reason that angles of 0 to 2π around two mutually perpendicular axes cover the sphere *twice* making such a putative co-ordinate system redundant and, incidentally, showing that there is no possibility of a non-redundant set of co-ordinates containing *two*, let alone three, canonical angular momentum components. In fact, once one component of the angular momentum "vector" has been chosen as conjugate to an angular variable this choice excludes other angular momentum components from being conjugate to any other angular co-ordinate in any co-ordinate system.

Some further insight into the problem may be obtained by considering ignorable co-ordinates - the starting point of the transformation theory. In Cartesians if

$$\frac{\partial L}{\partial x} = \frac{\partial L}{\partial y} = \frac{\partial L}{\partial z} = 0$$

this usually implies the potential is a constant (say zero) and the vanishing of the above derivatives implies (via the Lagrange equations) the constancy of the three conjugate momentum components and we interpret this constancy by saying that in the absence of a potential function 3-dimensional space is isotropic with respect to linear displacement. Now, in spherical polars a free-particle Lagrangian is

$$L = \frac{1}{2}m(\dot{r}^2 + r^2\dot{\theta}^2 + r^2\sin^2\theta\dot{\phi}^2)$$

and

$$\frac{\partial L}{\partial \phi} = 0$$

but

$$\frac{\partial L}{\partial \theta} \neq 0$$

indicating a major a-symmetry between the two angles in the co-ordinate system: ϕ has a privileged position. Once a given axis is chosen then the homogeneity of space for *rotations* is destroyed *by that choice*. Thus, caution is required if there is a tendency to make too much of certain apparent equivalences between translations and rotations: between angular and linear co-ordinates.

We have seen above that p_θ is not constant for a free particle but p_θ is conjugate to the angular co-ordinate θ and the canonical equations can be expressed in spherical polar co-ordinates. It should be clear now what the source of the confusions about angular momentum actually are: they are verbal rather than essential.

We have been discussing two quite distinct concepts and confusing them together because of similar terminology and certain intuitive expectations. These confusions are compounded by a *contingent* connection between the two which occurs in a familiar co-ordinate system. The clues to the resolution of the confusion lie in the fact that it is the *z-component* of $\vec{\ell}$ which is conjugate to ϕ (not z) and the fact that p_θ *is* a valid momentum component conjugate to θ . The problem is simply the *simultaneous existence* of two quite different quantities: the angular momentum "vector" and the momentum components conjugate to angular co-ordinates. These two quantities are defined independently of each other and, in general there will be no simple connection between them. In fact, as we have seen, there may be a *contingent* connection between them in the sense that (in spherical polar co-ordinates, at least) one of the *Cartesian* components of the angular momentum vector is identical to a momentum component conjugate to an angular variable. This is nothing more or less than a co-incidence which has, unfortunately, served to muddy the distinction between the two separate quantities. If one considers the 11 orthogonal co-ordinate systems in 3-dimensional space - many of which have dimensionless (angular) members - the role of θ in spherical polars is more typical. The momentum component $\partial L / \partial \dot{\theta}$ is a proper conjugate momentum component: conjugate, that is, to an angular variable but it is not a component of the angular momentum "vector", in particular it is not a Cartesian component of the angular momentum "vector".

Even if the position of a point cannot be specified by three angles it might be thought that the three Euler angles (for example) which are used to specify the orientation of a rotating body might past muster as "rotational canonical" co-ordinates. These three angles are, in fact, not sufficient to define the orientation of a rigid body with respect to a fixed "global" co-ordinate frame because one must specify the *order* in which the rotations are performed in fixing the body's orientation. Such a set of co-ordinates cannot be brought into the canonical formalism.

In the last few paragraphs, the word vector has been put in quotes when used in conjunction with angular momentum. This is deliberate and another attempt to draw attention to the difference between the angular momentum "vector" and momentum components conjugate to angular co-ordinates or, indeed, to any canonical momentum component. Angular momentum, defined as it is in terms of the vector product, is in fact a bivector or anti-symmetric second-rank tensor and it is only the co-incidence in 3-dimensional space that $3 = 3(3\text{-}1)/2$ which enables the components of this bivector to be put into one-one correspondence with the components of a vector.

This extended discussion of some of the properties and peculiarities of angular momentum was initiated by the more general question "under what conditions is a co-ordinate (or momentum component) suitable to be used in the canonical equations". Although angular variables and angular momentum provide special confusions and are the most celebrated case of "non-canonical" momenta it is time to return to the general case. However, it is worth remarking that these confusions between angular momenta and canonical momenta conjugate to angular co-ordinates, when taken over into quantum theory, are at the heart of the Einstein-Podolsky-Rosen (EPR) paradox.

The techniques of the Transformation theory are specially suited for this investigation since the only limitations placed on the allowed transformations are:

1. The co-ordinates should be a complete and non-redundant set: the non-vanishing of the Jacobian of the transformation.

2. Hamilton's canonical equations should have the same *form* in all allowed co-ordinate systems.

That is, the condition that a set of co-ordinates and momentum are canonical is "built into" the Transformation theory.

In section 2.4 it was shown that if the canonical equations were to have the same form in two different co-ordinate systems then there exists a function of the 12N+1 variables q^i, p_i, Q^j, P_j and (possibly) t from which the details of the transformation may be obtained. Crucial to this development is the fact that, of the $12N$ variables, only $6N$ may be independent: the other $6N$ are to be generated from these independent ones by the very transformation provided by this function. Now the choice of *which* $6N$ are chosen as independent variables is (formally if not practically) arbitrary and in 2.4 the most obvious choice was taken; writing the function which generates the transformation from q^i to Q^j as an explicit function of the q^i and Q^j . But there are three other obvious possibilities, if our original choice is $F_1(q^i, Q^j, t)$ then using the initial and final co-ordinates and momenta as units we may choose the independent variables in three other ways to define the functions

$$F_2(q^i, P_j, t) \quad : \quad F_3(p_i, Q^j, t)$$

and

$$F_4(p_i, P_j, t)$$

and all are related. Of course there is no reason why one should not choose *some* q^i and *some* p_i etc. but such choices have little theoretical or practical interest. We saw in 2.4 that, in addition to the sets (q^i, p_i) and (Q^j, P_j) satisfying the canonical equations which they were required to do, the p_i and P_j were related to F_1 by

$$p_i = \frac{\partial F_1}{\partial q^i} \quad : \quad P_j = -\frac{\partial F_1}{\partial Q^j}$$

That is

$$\frac{\partial p_i}{\partial Q^j} = \frac{\partial^2 F_1}{\partial Q^j \partial q^i} = -\frac{\partial P_j}{\partial q^i}$$

It is straightforward to show that similar relationships hold for the other choices:

$$\frac{\partial q^i}{\partial Q^j} = \frac{\partial P_j}{\partial p_i} \quad : \quad \frac{\partial q^i}{\partial P_j} = -\frac{\partial Q^j}{\partial p_i} \quad : \quad \frac{\partial p_i}{\partial P_j} = \frac{\partial Q^j}{\partial q^i} \qquad (2.6.3)$$

These four relationships fix the nature of the allowed transformations; that is, since the initial co-ordinates and momenta are arbitrary, they fix the possible co-ordinate systems in which Hamilton's canonical equations may be expressed, together with the requirement that the Jacobian of the transformation to a known set of complete and non-redundant set of co-ordinates be non-zero.

It is usual to express these relationships in a form which displays the *structure* of the results rather than its expression in terms of two co-ordinate systems although, of course, our co-ordinate systems are arbitrary.

The *Poisson Bracket* of X and Y (two quantities depending on co-ordinates, momenta and possibly time) is defined by:

$$[X, Y]_{q,p} = \sum_{i=1}^{3N} \left[\frac{\partial X}{\partial q^i} \frac{\partial Y}{\partial p_i} - \frac{\partial X}{\partial p_i} \frac{\partial Y}{\partial q^i} \right]$$

and, in particular.

$$[Q^i, P_j]_{q,p} = \sum_{k=1}^{3N} [\frac{\partial Q^i}{\partial q^k}\frac{\partial P_j}{\partial p_k} - \frac{\partial Q^i}{\partial p_k}\frac{\partial P_j}{\partial q^k}]$$

which, when the relationships 2.6.3 are used, becomes

$$[Q^i, P_j]_{q,p} = \frac{\partial Q^i}{\partial Q^j} = \delta_{i,j} = [Q^i, P_j]_{Q,P}$$

and similarly,

$$[Q^i, Q^j]_{q,p} = [Q^i, q^j]_{Q,P} = 0 \quad : \quad [P_i, P_j]_{q,p} = [P_i, P_j]_{Q,P} = 0$$

That is, the Poisson Bracket of the co-ordinates and momentum components are invariant with respect to *which* co-ordinate system they are evaluated in; so that these relationships which are derived using the relationships 2.6.3 may serve as indicators of allowed co-ordinates and conjugate momenta in the canonical equations.

It must be remembered, however, that there are *three* of them obtained from the four original restrictions on the q^i p_i Q^j P_j . One cannot make the judgement that a particular co-ordinate or, more important, a particular momentum is acceptable to the canonical formalism because it satisfies *just one* of these relationships among the Poisson Brackets.

Returning to angular momenta, we may use the convenience of the invariance of the Poisson brackets to evaluate them for the Cartesian components of angular momentum in Cartesian co-ordinates: obviously

$$[x, x] = [x, y] = ... = 0$$

But, for example

$$[\ell_x, \ell_y] = \ell_z$$

and

$$[x, \ell_y] = z$$

confirming the fact that the angular momentum components are not conjugate to a set of canonical co-ordinates. It is easy to show that

$$[\ell_x, \ell^2] = [\ell_y, \ell^2] = [\ell_z, \ell^2] = 0$$

but this does not admit ℓ^2 into the canonical scheme because there is no co-ordinate to which ℓ^2 is conjugate and so the other two relationships are *trivially* not satisfied. To emphasise the point once more, the components of the angular momentum "vector" and the scalar "square of the angular momentum" cannot be brought into the canonical formalism.

Chapter 3

Transition to Schrödinger's Mechanics

3.1 Introduction

We have seen in the last chapter that it is possible to compress all of the classical mechanics of a system of particles into a single partial differential equation in $3N+1$ variables from whose solution we can extract the trajectories of all the particles. Informal comments indicated that the function S contains a description of all possible trajectories of the particles consistent with Newton's Laws and that a new point of view was incipiently possible: that the referent of the Hamilton-Jacobi equation is an *ensemble* of systems of particles whose motions are all governed by Newton's Laws and which differ by their initial conditions. That is, we emphasise the fact that S contains information about the relationships between the initial conditions as well as providing (e.g) the momentum as a function of space and time.

It is worthwhile pointing out the difference between the dependence of S on the q^i, for example, and the dependence of H (or L) on the q^i. The S which solves the Hamilton-Jacobi equation is defined for all values of the q^i (and possibly t) just as H and L are. But S at every point satisfies the dynamical law; i.e. at every point $S(q^i)$ and $\partial S/\partial q^i$ refer only to real (allowed) motions of the particles. In the case of H and L their dependence on the q^i is much wider than this; only when the q^i in H (or L) satisfy the equations arising from the variational principle (the canonical equations) do the q^i become allowed trajectories.

Thus, speaking coloquially, we might say that "S fills out configuration space with allowed trajectories"so that this function, unlike H or L has an objectively real referent for each and every value of its arguments q^i and t. What we are saying here is that, by a shift of emphasis, we may regard this referent as an ensemble of all the possible motions of the system of particles "co-existing"rather than S having the referent of all the potential motions of a single system of particles.

This shift of emphasis is made because the next development in the mechanics of systems of particles came in attempts to describe the motions of particles for which there was (and is) no hope of being able to specify the initial conditions: atomic and sub-atomic particles. For all practical purposes, experiments can neither determine nor infer the "initial"conditions of, for example, the motion of the electrons in the carbon atom so that any mechanics which requires the specification of such data is doomed to impotence in the sub-atomic domain. At the turn of the century the pressure to develop a system of dynamics to account for the bewildering array of experimental discoveries was acute and all the best minds in theoretical physics were concentrated on this problem.

Schrödinger took the Hamilton-Jacobi equation as his starting point for the creation of a new system of particle mechanics valid in the region of the very small and the very light. His contribution in his epoch-making first paper may be summarised in two steps; one apparently trivial and one boldly new. We will examine them in sequence beginning with the smaller of the two steps. In looking at Schrödinger's work we must, of course, guard against the idea that his mechanics can be "derived"from the Hamilton-Jacobi equation: it can't. Schrödinger's mechanics is a new creation, it contains new intuition about reality which mathematical manipulation can never supply: we can only hope to make the translation a little smoother, to bring classical particle mechanics and quantum mechanics close together before making a jump. The difference between pedagogy and creation is that, when *we* jump, we know that there is something there to jump to.

3.2   A New Notation for Action

The solution of the Hamilton-Jacobi equation, S, has the dimensions of *energy* × *time* ("action") and its time derivatives have, as we have seen, the dimensions of energy and, in particular, $-\partial S/\partial t$ is the total energy of the system as a function of the q^i (and possibly t).

Schrödinger, in his investigations of the Hamilton- Jacobi equation made an apparently trivial change of notation writing:

$$S = K \ln \psi \quad : \quad \psi = \exp(S/K) \quad\quad (3.2.1)$$

where the numerical factor was added since $\ln \psi$ has no dimensions and

S should have dimensions of action; obviously K depends on the system of units being employed for actual calculations: a "natural" choice of units would be one which gave K the numerical value of unity.

If this substitution is made in the Hamilton- Jacobi equation we need the relationship

$$\frac{\partial S}{\partial q^i} = \frac{K}{\psi}\frac{\partial \psi}{\partial q^i} \qquad (3.2.2)$$

i.e.

$$\frac{\partial \psi}{\partial q^i} = \frac{\psi}{K}\frac{\partial S}{\partial q^i}$$

for the momenta, and then the equation becomes

$$H(q^i, \frac{K}{\psi}\frac{\partial \psi}{\partial q^i}, t) = \frac{K}{\psi}\frac{\partial \psi}{\partial t}$$

which is the Hamilton-Jacobi equation in the new notation. However, this "mere notational change" provides an additional motivation for the change of emphasis in the interpretation of the equation which was mentioned in the previous chapter. For simplicity, consider the one-dimensional case with Hamiltonian

$$H(q,p) = \frac{1}{2m}p^2 + V(q)$$

H is independent of time so the Hamilton-Jacobi equation is

$$H(q, \frac{\partial S}{\partial q}) = E \qquad (3.2.4)$$

(where partial derivative notation has been retained in this one-dimensional case for consistency with the general case). Using Schrödinger's notation this equation becomes

$$H(q, \frac{K}{\psi}\frac{\partial \psi}{\partial q}) = E$$

i.e.

$$\frac{1}{2m}\frac{K^2}{\psi^2}(\frac{\partial \psi}{\partial q})^2 + V(q) = E$$

or

$$\frac{K^2}{2m}(\frac{\partial \psi}{\partial q})^2 + V(q)\psi^2 = E\psi^2 \qquad (3.2.5)$$

Now if we transform just the derivative part of the equation back into the original notation involving S using 3.2.2 we have:

$$\left(\frac{\partial \psi}{\partial q}\right)^2 = \frac{\psi^2}{K^2}\left(\frac{\partial S}{\partial q}\right)^2$$

i.e.

$$(\psi^2)\frac{1}{2m}\left(\frac{\partial S}{\partial q}\right)^2 + (\psi^2)V(q) = (\psi^2)E \qquad (3.2.6)$$

which is just a *multiple* of the Hamilton-Jacobi equation.

Of course, ψ^2 may be simply cancelled from 3.2.6 to emphasise that the Hamilton-Jacobi equation is recoverable from a change of notation! But suppose now that we insist that 3.2.6 *be* the Hamilton-Jacobi equation i.e. without cancellation. We must require $\psi^2 = 1$ which is a contradiction since ψ is not a constant if S is not. If we allow *complex* ψ , going through the whole procedure again generates

$$|\psi|^2\frac{1}{2m}\left(\frac{\partial S}{\partial q}\right)^2 + |\psi|^2 V(q) = |\psi|^2 E \qquad (3.2.7)$$

and insisting that 3.2.7 be identical to the Hamilton-Jacobi equation yields $K = \pm ik$ (say)[1]

$$S = -ik\ln\psi \;\; : \;\; \psi = \exp(-S/ik)$$

$$\psi = \exp(iS/k)$$

where k and S are real and the minus sign has been chosen for conventional reasons.

Now we can attempt to interpret 3.2.7. Let us do this by temporarily ignoring the fact that $|\psi^2|$ is constrained to be unity and look at the *form* of 3.2.7 when translated back into co-ordinate and momentum variables, i.e.

$$|\psi|^2\frac{1}{2m}p^2 + |\psi|^2 V(q) = |\psi|^2 E \qquad (3.2.8)$$

Taking the first term, it has the form

$$(positive function) \times (Kinetic Energy)$$

that is, it has the form of a *distribution* of kinetic energy. Similarly, the other two terms in 3.2.8 have the form of a distribution of potential energy and a distribution of total energy, respectively. Further, the function $|\psi|^2$ has some of the properties of a distribution function: it is always positive and it is bounded. Recall again that it is possible to interpret S as referring to an ensemble of systems and we have the beginnings of a new approach: perhaps an equation reminiscent of the Hamilton-Jacobi equation can be made to yield information about the way in which the allowed trajectories

---

[1] This is where we depart from Schrödinger's own derivation slightly; Schrödinger had K real.

contain in S are distributed in space *whatever the initial conditions* of the motion.

With these hopeful ideas in mind we now go back to equation 3.2.1 again and *allow S to be complex* while retaining the same relationship between S and ψ . This is now not a trivial change of notation generating mere tautologies; it is a new assumption in mechanics since classical mechanics has no use for a two-component S function, all the dynamics comes out of a real S. The whole object of this change is to introduce a "new degree of freedom"into the development which will separate $|\psi|^2$ from S and the momentum in order that $|\psi|^2$ can play the role of a genuine distribution function not constrained to be a constant.

Thus, by writing

$$S - iR = -ik \ln \psi \quad : \quad \psi = \exp[(R + iS)/k] \qquad (3.2.9)$$

we have

$$|\psi|^2 = \exp(2R/k) = \rho(q) \quad \text{(say)}$$

and equation 3.2.8 becomes

$$\rho \frac{1}{2m} p^2 + \rho V(q) = \rho E \qquad (3.2.10)$$

where ρ is a function of space and a discussion of the meaning of p (the momentum) has been deliberately deferred. We can now use the same interpretation of the terms in 3.2.10 as before: each term is a distribution function multiplying an energy function. But what is $\rho(q)$ a distribution *of* and how is it to be determined? We have introduced a new function R and no equation to determine it: we can still formally cancel ρ from 3.2.10 leaving the Hamilton-Jacobi equation for S. This equation generates all possible trajectories for a given $V(q)$ and there is no reason in classical mechanics to look for a distribution of these trajectories; for classical mechanics is only interested in the trajectories of particular particles and is indifferent to their distribution in space.

But Schrödinger was acutely aware of the need to develop a new mechanics of sub-atomic particles and, no doubt, even more acutely aware that a new mechanics cannot be got by changes in notation, however suggestive those changes might be. What was needed to enable the theoretical understanding of the dynamics of sub-atomic particles was an equation which would "contain"or "go over into"the Hamilton-Jacobi equation for large masses and allow for our ignorance of "initial conditions"in the sub-atomic world. As it turned out, the combination of the Hamilton-Jacobi equation and an equation for ρ was just what was needed. Schrödinger was able to present a *single* equation which generated both R and S, the distribution and the action.

Historically Schrödinger's "reasoning"(if creative thinking can be called reasoning) was different from the development given here and was based on an analogy between optics and mechanics originally due to Hamilton. This very analogy led to some confusion about the interpretation of quantum mechanics which we are trying to side-step here: as we remarked above pedagogy is not creativity.

3.3   Schrödinger's Dynamical Law

We have tried to present a plausible way in which the Hamilton-Jacobi equation could be extended and generalised. The elements are:

1. The concentration, in classical mechanics, of information about trajectories into a single scalar function S

2. The function S being regarded as referring to an ensemble of systems differing in their initial conditions.

3. The possibility of relating S to a distribution function by the (formally) slight generalisation of admitting complex S.

We use the distribution function ρ to define "densities"which will be put on a firmer basis later in this chapter. In the Hamilton-Jacobi equation, each term in the Hamiltonian is replaced by a "corresponding"density. Again using a one-particle system for simplicity, the Hamilton-Jacobi equation has $\partial S/\partial q^i$ as the momentum, and, since we are dealing with a one-particle system, $\partial S/\partial q^i$ is also the momentum *per particle*. We take this idea over into the Schrödinger theory (with the generalisation of complex S), using the kinetic, potential and total energy *per particle*. Thus

$$H(q^i, \frac{\partial S}{\partial q^i}, t)$$

is the Hamiltonian per particle (in a one-particle system) and

$$-\frac{\partial S}{\partial t}$$

is the energy per particle.

In Schrödinger's theory, multiplying these energy quantities by the particle distribution function $\rho(q^i, t)$ generates the actual densities; e.g. the Hamiltonian density

$$d_H = \rho H(q^i, \frac{\partial S}{\partial q^i}, t)$$

and the energy density

$$d_E = -\rho \frac{\partial S}{\partial t}$$

In Schrödinger's notation these important densities become

$$d_H = |\psi|^2 H(q^i, \frac{-ik}{\psi} \frac{\partial \psi}{\partial q^j}, t)$$

and

$$d_E = |\psi|^2 \frac{ik}{\psi} \frac{\partial \psi}{\partial t}$$

What is now required is an "equation of motion"- a new dynamical law - which *replaces* the Hamilton-Jacobi equation in these new circumstances.

The Hamilton-Jacobi equation can be given a quite simple verbal formulation:

> Of all the possible trajectories $q^i(t)$ and momenta $p_i(t)$ of the particles described by $H(q^i, p_i, t)$, the ones which occur in nature are those for which the value of the function H is numerically equal to the energy of the system.

That is, for real motions of the particles in the system, only those q^i and p_i are allowed in H which make

$$H = -\frac{\partial S}{\partial t} = E \quad (\text{say})$$

We have been at pains to stress earlier that the Hamiltonian *function* has, as its domain, all possible q^i and p_i (and possibly t) whether or not they meet the requirements of Newton's Laws via the canonical equations.

Schrödinger replaces this requirement with a less stringent constraint which is just as easy to state verbally:

> Of all possible trajectories q^i and momenta p_i of the particles described by H the ones which occur in nature are the ones which *on the average* over space and time make H equal to the energy of the system.

That is, Schrödinger's modification of the Hamilton-Jacobi principle is that the Hamilton-Jacobi equation does not have to be obeyed point by point in a configuration space of $3N$ dimensions but only in the mean over all space.

In symbols this new dynamical law becomes:

$$\int d_H dV dt = \int d_E dV dt \qquad (3.3.1)$$

where the Hamiltonian and energy densities d_H and d_E are defined above. In view of the generalisation of "action" by Schrödinger to include the complex case, it is perhaps worth introducing a slight change of notation. We will write the complex "action" as:

$$S' = S - iR$$

retaining S for the classical action and retaining the symbol S in the new notation S' as a reminder of the pedigree of Schrödinger's new function. This means that, in full,

$$d_H = \rho H(q^i, \frac{\partial S'}{\partial q^i}, t) \qquad (3.3.2)$$

and

$$d_E = -\rho \frac{\partial S'}{\partial t}$$

In terms of Schrödinger's ψ notation, which will be used the more frequently of the two, the densities are

$$d_H = |\psi|^2 H(q^j, -\frac{ik}{\psi}\frac{\partial \psi}{\partial q^j}, t)$$

of course, 3.3.1 may be written

$$\int (d_H - d_E)dV dt = constant \qquad (3.3.3)$$

or

$$\delta \int (d_H - d_E) = 0$$

which is a variational principle from which the equation of motion - "Schrödinger's Equation" may be derived. This condition, which generates an equation for (complex) ψ, is sufficient to determine R and S, i.e. R the distribution function and S which provides information about the momentum averages.

This is Schrödinger's new dynamical law and will be called, from now on, the *Schrödinger Condition*.

Application of standard variational methods to 3.3.3 generates the Schrödinger equation for ψ and some boundary conditions which will need elucidation.

But, before going into any technical details of the generation and solution of the Schrödinger equation and any problems of interpretation which that might bring, we can anticipate some of the problems by referring back to some properties of the solution of the Hamilton-Jacobi equation S. It has been noted that the function S contains all possible trajectories and, although this was not stressed, what is more it contains them all just once; for given values of the constants of integration ("initial conditions") there is just one solution of the Hamilton-Jacobi equation.[2] Mathematically, the solution S "fills" the configuration space. Thus in the extended, complex, S' the function $|\psi|^2$ can be expected to contain reference to all possible trajectories with equal weights. For example, for an isolated single particle moving in a potential V the distribution is over all possible trajectories with constant energy obeying 3.3.3: the trajectories only differ in "initial" conditions. Now in studying a dynamical law which only determines the properties of *averages* of dynamical variables (H and E) over trajectories we should not be surprised if we cannot recover the *individual* trajectories over which the averages have been taken. Schrödinger's law generates an equation for the distributions of particles in space averaged over *all possible* initial conditions i.e. over all possible trajectories. The individual trajectories are not even required to solve an Hamilton-Jacobi equation (or, indeed any equation) that is, individually, the particle trajectories are not required to solve Newton's law, only the averages are fixed by 3.3.3.

That is not to say that the particles in each member of the ensemble do not have perfectly definite trajectories along which some (as yet unknown) laws are obeyed, it is simply that condition 3.3.3 *does not tell us what these trajectories are.* For example, in the classical mechanics of an isolated system we have at every point in space

$$H - E = 0$$

But all that 3.3.3 requires is that if

$$H - E = \delta$$

for some region of space, then there are enough allowed trajectories so that, for some other region of space,

$$H - E = -\delta$$

---

[2] Here we need some suitable limits on cyclic co-ordinates to avoid covering space several times

so that, on average over all space:

$$< H > - < E > = 0$$

The averages are able to be cast into a form reminiscent of Newton's law — the Ehrenfest relationships — but again only the averages obey these relationships not the individual trajectories.

The difference between the mechanics generated by 3.3.3 and classical statistical mechanics which also deals with ensembles, distributions and ensemble averages will help to make the above points clearer. In quantum mechanics the individuals comprising the ensemble differ by initial conditions and are required to solve the Hamilton-Jacobi on average over space and time; individual members of the ensemble are allowed to have motions which do not obey the Hamilton-Jacobi equation. In classical statistical mechanics the motion of *each member* of the ensemble is required to solve the Hamilton-Jacobi equation exactly and averaging is done with these exact solutions.

That is, the condition for the satisfaction of the relevant dynamical law is:

$$H - E = 0 \quad \text{Classical Particle Mechanics}$$
$$\rho = \delta(H - E) \quad \text{Classical Statistical Mechanics}$$
$$< H - E > = 0 \quad \text{Schrödinger's Mechanics}$$

Showing the clear "mean value" nature of the quantum case.

In the next section a simplified example is used to attempt to clarify the nature of the alleged distribution function $|\psi|^2$ and also to clear up the relationship between a complex "action" function $S - iR$ and momentum components, something outstanding from the introduction of equation 3.2.9.

3.4   Densities and Momenta

In this section attention is restricted to ensembles of single particle systems in ordinary 3-dimensional space for simplicity and intuitive appeal since it is intended to try to picture the various qualities appearing in Schrödinger's theory. The function ψ fixed by the variational requirement 3.3.3 is, in general, complex and its primary physical interpretation is via $|\psi|^2$ as indicated in the build-up to 3.3.3. For an isolated (constant energy) system with a time-independent Hamiltonian function, $-\partial S/\partial t = constant = E$ (say) and so the function ψ is dependent on time only through an exponential factor of modulus unity $(\exp(iEt/k))$ which does not appear in $|\psi|^2$ and so we neglect it for the moment. The interpretation of $|\psi|^2$ as a particle distribution over possible trajectories is clearly that

$$\int_W |\psi|^2 dV \quad W \subset R^3 \tag{3.4.1}$$

is related to the number of members of the ensemble which have a particle in the region W of three-dimensional space. Equally clearly, since we are dealing with *infinite* numbers of trajectories, numbers like 3.4.1 must be judged *relatively*, that is judged by reference to the size of

$$\int |\psi|^2 dV \tag{3.4.2}$$

over all space. If 3.4.2 is finite (and this is not always the case) then ψ can be re-scaled by a constant factor so that 3.4.2 has some convenient value : unity or the number of particles in a member of the ensemble are obvious choices. In fact, it is convenient to normalise ψ to unity i.e. insist, by use of a numerical factor that

$$\int |\psi|^2 dV = 1 \tag{3.4.3}$$

so that the relative numbers

$$\frac{\int_W |\psi|^2 dV}{\int |\psi|^2 dV} \tag{3.4.4}$$

are the same as the numbers 3.4.1

Now if 3.4.3 is imposed then the measures 3.4.1 satisfy Kolmogorov's axioms for an uninterpreted probability system, for if:

$$P(W) = \int_W |\psi|^2 dV$$

then

- $P(W) \geq P(W')$ if $W \supset W'$

- $P(W_1) + P(W_2) = P(W_1 + W_2)$ if $W_1 \cap W_2 = \emptyset$

- $P(R^3) = 1$

and $|\psi|^2$ is a distribution function since

- $|\psi|^2$ is single valued

- $|\psi|^2$ is continuous

- $|\psi|^2 \geq 0$

We may therefore use any or all of the techniques and concepts of probability theory. In particular we may refer to the numbers 3.4.4 (normalised measures) as "the probability that a particle be in region W"(recall, for the moment we are concentrating on single particle system for simplicity). That is, for pictorial purposes, we may concentrate attention on a "typical"member of the ensemble and use probability statements about it. This change of viewpoint carries its own pitfalls as we shall see later. [3]

In this spirit we may interpret $|\psi|^2$ as the distribution of "the"particle in the typical ensemble member of one particle systems or even $|\psi|^2$ as the "particle density"in this fictitious typical system. Obviously, this latter kind of language becomes more useful and realistic in systems of many particles; it makes more sense, for example, to speak of the "electron density in the uranium atom"than of the "election density in the hydrogen atom"but both are shorthand for "the relative number of members of an ensemble of atoms with at least one electron in the given volume per unit volume".

The extension of these ideas to ensembles of many-particle systems is straightforward and we may, with caution, use the probability concept and say that

$$\int_{W_1, W_2, \ldots} |\psi(q^1, q^2, \ldots, q^3 N)|^2 dV_1 dV_2 \ldots dV_N$$

is the probability that particle 1 is in region W_1 , particle 2 in region W_2 etc. Always provided that the function ψ has been normalised to unity in $3N$-dimensional space.

We must now address a problem of interpretation which is long overdue; having been ignored in the combination of analogy and formal manipulations of the last section. In Hamilton-Jacobi theory the momenta are generated from S by partial differentiation:

$$p_i = \frac{\partial S}{\partial q^i}$$

and in the one-particle case these momenta are, of course, the momenta per particle. That is, before allowing S to be complex, in Schrödinger's

---

[3] One has to be prepared for some surprising properties of such "typical"members. The average family in the U.K has 2.3 children, for example and the average of dice throws is 3.5; neither of these "typical"results occur in any real family or dice throw.

notation the classical momentum per particle is

$$p_i = -\frac{ik}{\psi}\frac{\partial\psi}{\partial q^i}$$

In fact, as we shall see, this relationship is unchanged in Schrödinger's quantum theory. But in the last section we allowed S to be complex *without* any consideration of the effect that this new freedom might have on the validity of the old relationship between S and the momentum components. That is, what is the physical interpretation when

$$\frac{\partial S}{\partial q^k} \to \frac{\partial S}{\partial q^k} - i\frac{\partial R}{\partial q^k}$$

in particular, what is the interpretation of

$$\frac{\partial R}{\partial q^k} \ ?$$

Writing out ψ in its full "action-exponential" form allows some clarification:

$$\psi = \exp(R + iS) = \exp(R)\exp(iS)$$
$$-i\frac{\partial\psi}{\partial q} = \exp(R)\exp(iS)\left(-i\frac{\partial R}{\partial q} + \frac{\partial R}{\partial q}\right)$$

and so

$$\psi^*(-i\frac{\partial\psi}{\partial q}) = -i\exp(R)\frac{\partial}{\partial q}\exp(R) + \exp(2R)\frac{\partial S}{\partial q}$$

Now the particle distribution is given by

$$|\psi|^2 = \exp(2R)$$

so that the second of the above two terms (the real component, arising from the real action) is just analogous to a classical distribution of momentum. The first term, wholly imaginary, is new; representing contributions to the momentum density due to the gradient of the particle *distribution*.

If ψ is normalised to unity the *mean value* of the momentum associated with ψ is just

$$\int_a^b \psi^*\left(-i\frac{\partial\psi}{\partial q}\right) dq = -i\left[\frac{1}{2}\exp(2R)\right]_a^b + \int_a^b \exp(2R)\frac{\partial S}{\partial q}dq$$

Now the first of these two contributions to the mean value arises from the boundaries of the particle distribution which, as we shall see later when we develop an equation satisfied by ψ, always vanishes.[4] Thus the imaginary part of the quantum action makes no contribution to the *mean value* of the momentum components but does make a contribution to the momentum distribution.

What is the meaning of a contribution to a momentum component due to changes in particle distribution?

The best way to answer this question - which is a substantive, physical question not a mathematical one and so is impervious to further manipulations — is by consideration of a concrete example: an attempt at the visualisation of an ensemble for a particular, simple case. We will consider angular momentum as a model for momenta simply because, in imagination, we can stay in one place and observe particles with angular momentum about that place.

Suppose then we have an ensemble of one-particle systems each of which has its particle in an orbit around a fixed point and suppose that the whole ensemble has some average angular momentum. The members of the ensemble differ, as usual, by their initial conditions. In order to visualise this ensemble let us now imagine that all the particles are in the same physical space (i.e. not inhabiting their own copy of space as they do in the ensemble) but non-interacting: point particles which do not collide or have any interactions of any kind, say. Initially we assume that they all have the same angular momentum (which is then the average value) but that they are distributed at random in space (a subset of all possible initial positions). From the viewpoint of the fixed "centre" of their motion we would see them all rotating with fixed inter-particle distances: rotating as a solid body. They would rotate as the stars appear to do around the earth: as if fixed on a crystal sphere rotating with constant velocity.

Now suppose, while retaining the same average angular momentum, the members of the ensemble do not all have the same angular momentum (angular velocity) [5] and let us see how the same visualisation of the particles as "stars" in the night sky is affected. What would be seen would depend on the size of the deviations of the individual's angular momentum from the average. Let us assume for simplicity that only a minority of the individuals have angular momenta different from the average and that their deviations from the average are small compared to that average. This time, looking out from the centre we could easily pick out the non-constant angular momenta as "stars" which take part in the overall rotation but which also wandered about in the sky: their *relative* positions do not remain fixed once their

---

[4] This class of $\psi's$ is just the one which makes the momentum operators Hermitian

[5] We have used a subset of these members of constant angular momentum consisting of those with constant radius from the centre - the argument is unaffected.

angular momenta are not constant. In a word, those particles with non-constant angular momenta would appear as *planets* ("wanderers").

With the aid of these preliminary simplified cases we can now imagine the general case in which each member of the ensemble (each of our non-interacting particles: "stars") has its own angular momentum (velocity) which differs from the average by a greater or lesser amount. If the deviations from the average are small then the picture of the total angular momentum of the ensemble is one in which there is still a perceptible overall turning (daily rotation in the cosmic analogy) with each particle making its own apparent motions about some mean position. At any one instant in time there will be many particles moving in the direction of the "overall turning"- the direction of the average - but also a number stationary or moving in the opposite direction. As the deviations from the average become larger, the motion will appear more and more chaotic to a central (earthbound) observer but the ideas are still the same.

Now to the point at issue - consider how this picture of the angular momenta of a set of randomly distributed non-interacting particles is related to the *angular distribution* of those particles. We can get an idea of the angular distribution (distribution in a co-ordinate conjugate to the momentum component) by counting the number of particles passing a particular line-of-sight per unit time. If the ensemble is angular-momentum homogeneous as in our very first example and, if the particles are random and numerous enough, the number of particles passing a particular line-of-sight per unit time is *constant*. However in the second case, if some particles' angular momenta deviate from the average then, in a given interval of time, such a particle may cross and re-cross (in the opposite direction) a given line-of-sight, and, depending on the details of the motion, such processes may increase or decrease the particle count per unit time.

Thus, *deviations* from an angular-momentum-homogeneous ensemble "cause" deviations from a constant particle count in a conjugate spatial co-ordinate. And so, at any point, non-uniformity in the spatial distribution of such particles in a given co-ordinate (if there are enough of them) is diagnostic for non-constancy of the conjugate momentum component or, more precisely, the momentum non-homogeneity of the ensemble. That is to say there is not a direct correlation between the spatial distribution of an ensemble of particles and the momenta of these particles but there is a direct correlation between *changes* in such a distribution in a given co-ordinate and *deviations* from the mean in the conjugate momentum component. This conclusion looks optimistic: it means that there is a necessary connection between gradients of $|\psi|^2$ (i.e. of $\exp(2R)$) and the

conjugate momenta.[6] Angular momenta have been used in the above pictorial development because of the familiar "stellar" analogy but the thread
of the argument is not restricted to angular momentum components. One
can almost as easily visualise counting experiments being used to observe
deviations from a stream of steadily-flowing particles with some average
linear momentum. Anything which obstructs or helps the linear motion
(like a source of potential) causes deviations from constant linear motion
and therefore a change in the spatial distribution of the particles in the
conjugate (Cartesian) co-ordinate.

We have the main conclusion then that in allowing complex S' and
simultaneously requiring the Hamilton-Jacobi equation to be satisfied on
the average only means that, in addition to a relationship existing between
gradients of the original (real) action and momenta:

$$p_k = \frac{\partial S}{\partial q^k}$$

there is a relationship between gradients of the imaginary part of the action
and deviations from the average momentum components.

In the classical Hamilton-Jacobi theory the function S contains information about the (constant) initial momentum components in that the
derivatives of S with respect to the integration constants (the initial co-
ordinates) are just these constant momentum components. This was the
import of the approach to the Hamilton-Jacobi equation through the classical transformation theory: section 2.4. Now one can immediately see that
in certain well-chosen co-ordinate systems the initial constant momentum
components will be unchanged throughout the motion: cylindrically symmetrical potentials will not attenuate an angular momentum component
about the cylinder axis, for example. That is the gradients of the (real)
classical action can, in some cases, be made to yield the constant conjugate
momentum components. In the case of Schrödinger's theory if the imaginary part of its action is zero for some co-ordinate, the spatial distribution
in that co-ordinate is constant and the momentum component of all the
members of the ensemble is therefore the same: there are no deviations.
We are now ready to make an assumption about the relationship of the
quantum momentum components to the complex action in the Schrödinger
theory.

The gradients of the real part of the complex action are proportional to
the *average* momentum components per particle of the ensemble while the

---

[6] The detailed form of this connection will be assumed in what follows but a more thor-
ough investigation would yield the correct formula (to within a multiplicative constant).

gradients of the imaginary part of the complex action are proportional to the *deviations* from the average of the momentum components per particle. Thus the total momentum distribution per particle at any point in space is proportional to

$$\frac{\partial S'}{\partial q^j} = \frac{\partial S}{\partial q^j} - i\frac{\partial R}{\partial q^j} = -\frac{ik}{\psi}\frac{\partial \psi}{\partial q^j}$$

as we implicitly assumed earlier without justification Since the particle distribution is given by $|\psi|^2$ we can easily evaluate the total momentum distribution of the whole, ensemble it is just the particle distribution multiplied by the momentum per particle:

$$|\psi|^2 \left(-\frac{ik}{\psi}\frac{\partial \psi}{\partial q^j}\right) = \psi^*\psi\left(-\frac{ik}{\psi}\frac{\partial \psi}{\partial q^j}\right)$$

$$= \psi^*\left(-ik\frac{\partial \psi}{\partial q^j}\right)$$

conjugate to co-ordinate q^j.

In physical (pictorial) terms this can be interpreted to mean that in quantum mechanics the real part of the function S' (S) describes the momenta of the ensemble *as a whole*, (the daily rotation in our analogy) while the imaginary part of S' (R) describes the spatial distributions and those momenta which cannot be reduced to motion of the ensemble as a whole by a suitable choice of viewpoint. If we had chosen to do our particle counts in the stellar analogy along a tangent instead of along a radius we would have obviously have obtained different results (corresponding to linear momentum) but by a suitable choice of point of view (co-ordinate system) we can extract the momentum component which has an average and deviation of the kind we have discussed. In short the division of S' into a real part and a imaginary part is unique but the choice of co-ordinate system in which to express ψ is not. We shall see in a later chapter that there is a systematic way of choosing the most appropriate co-ordinate system for a given potential function which reveals the division of S' into real and imaginary parts to best advantage.

The *need* for a complex function is now obvious, one needs two real functions to generate the two "degrees of freedom" in the ensemble model. Since the Schrödinger theory only deals with *averages* of dynamical qualities like momentum; and momenta appear squared in the Hamiltonian formulation then there is immediately the problem that the average of squares of quantities is not the same as the square of the averages if there are deviations from those averages. In terms of the analogy used earlier we may imagine an ensemble composed of equal numbers of of clockwise and anti-clockwise rotating particles of the same magnitude giving an ensemble average angular momentum of zero. But the mean angular kinetic energy - the square of the angular momenta - is not zero, it is certainly not the square of the mean angular momentum. In any theory dealing with distributions there must be provision for solving problems of this kind; in particular in quantum mechanics we must have ways of obtaining means and deviations from those means in order to obtain the distribution of other dynamical variables. In the above example, the mean angular momentum component is zero so that the real part of the complex S' is identically zero and only the imaginary part (which gives the deviations) remains: all the angular kinetic energy is due to ensemble members which have angular momentum different from the average (obviously - the average is zero!)

We are now in a position to summarise the physical interpretation of the complex action function introduced by Schrödinger by way of

$$\psi = -ik \ln S' = -ik \ln(S - iR)$$

- The "position density" is given by

$$|\psi|^2 = \exp(2R/k)$$

  in the sense that

$$\int_W |\psi|^2 dV$$

  is the probability that the particle be in a region W (for a single particle system with obvious extensions for many- particle systems). More explicitly, this number is the relative number of members of the ensemble which have a particle in region W

- The function

$$-\frac{ik}{\psi} \frac{\partial \psi}{\partial q^j}$$

  is the momentum component density conjugate to q^j per particle and therefore

$$\psi^* \left(-ik \frac{\partial \psi}{\partial q^j} \right)$$

  is the total momentum component density conjugate to q^j in the sense that

$$\int_W \psi^* \left(-ik \frac{\partial \psi}{\partial q^j} \right) dV$$

  is the "amount of momentum" in region W. More explicity, it is the sum of the momenta of the relative number of particles in region W given by the first expression.

With these interpretations we can set up expressions for the kinetic energy and potential energies in terms of ψ if we know the metric properties of the co-ordinate system used and therefore the relationship between the momenta and the (classical) kinetic energy expression. In Cartesian co-ordinates a typical contribution to T will be:

$$T = \frac{(particle\ density)}{2m} \times (momentum\ per\ particle)^2$$

$$= \frac{|\psi|^2}{2m} \left| -\frac{ik}{\psi} \frac{\partial \psi}{\partial x} \right|^2$$

$$= \frac{k^2}{2m} \left(\frac{\partial \psi^*}{\partial x} \right) \left(\frac{\partial \psi}{\partial x} \right)$$

i.e.

$$T = \frac{k^2}{2m} \left| \frac{\partial \psi}{\partial x} \right|^2$$

and

$$V = V(q^k)|\psi|^2 = V(q^k)\psi^*\psi$$

Before leaving the interpretation of ψ a word is necessary about the way in which the real and imaginary parts of the complex action were introduced. Since

$$\psi = -ik \ln(S - iR)$$

i.e. $\psi = \exp((R + iS)/k) = \exp(S'/k)$ the real (classical) action S appears in the wave function ψ multiplied by i and the "imaginary"(additional quantum) action appears not explicitly multiplied by i in a contrary way. The function ψ will be used more and more from now on but the function S' gives a cleaner separation of the classical and quantum components of the extended action and so will be used from time to time when it is necessary to draw attention specifically to this distinction. The function ψ does not display the separation of the "classical"and "quantum"components so cleanly.

3.5 The Schrödinger Equation

We can now turn to the use of standard variational calculus on equation 3.3.3 to derive an equation which determines ψ . The constant k which has appeared throughout the chapter relating ψ to the action S and S' is not to be found from theoretical considerations since it depends both on experiment and on the system of units employed: we shall choose to work in a "natural"system of units in which k has the numerical value unity (while still having the dimensions of action) and so disappears from our equations. We shall have reason to return to considerations of the *size* of k when we consider the limiting forms of quantum theory and the transition between the areas of applicability of classical and quantum mechanics. This decision implies that we take Planck's constant divided by 2π ($h/2\pi$), as a unit of action.

Schrödinger's requirement that the Hamiltonian function be equal to the energy an average has the general form of a variational principle of the type

$$\delta \mathcal{A} = 0$$

where

$$\mathcal{A} = \int A(\phi, \phi^*, \partial_i\phi, \partial_i\phi^*, \partial_t\phi, \partial_t\phi^*)dVdt$$

where ϕ is a function of the q^i and t and

$$\partial_i = \frac{\partial}{\partial q^i} \; ; \partial_t = \frac{\partial}{\partial t}$$

$$dV = \sqrt{g}dQ = \sqrt{g}\prod dq^i$$

where, as usual, g is the metric determinant of the co-ordinates q^i which may be evaluated as the Jacobian of the transformation between Cartesians and the q^i which we assume to be non-zero. For convenience of manipulation and visualisation let us work with a single particle in ordinary three-dimensional space i.e. configuration space is ordinary, real, space. In orthogonal co-ordinates for example $\sqrt{g} = h_1h_2h_3$ where the h_i are the scale functions associated with the tangent space basis in the usual way.

Thus $\mathcal{A}$ is a functional of ϕ and ϕ^* and it is desired to find a solution of the variational problem by choice of optimum ϕ; the standard problems in variational calculus. We use elementary methods. Let $\delta\phi$ and $\delta\phi^*$ be linearly independent variations in the linearly independent functions ϕ and ϕ^* so that we may investigate the variation in the functional $\mathcal{A}$ in the neighbourhood of its value at ϕ, ϕ^* by writing

$$\phi \to \phi + \delta\phi = \phi + \epsilon\eta$$
$$\phi^* \to \phi^* + \delta\phi^* = \phi^* + \epsilon\eta^*$$

where ϵ is a "small" real parameter and η, η^* are linearly independent functions arbitrary apart possibly from some boundary conditions. It is assumed that the integrand A is a sufficiently smooth function of ϕ and ϕ^*

so that it may be expanded as a Taylor series about ϕ, ϕ^*:

$$\delta A = \frac{\partial A}{\partial \phi}\delta\phi + \sum_{i=1}^{3N}\frac{\partial A}{\partial(\partial_i\phi)}\delta(\partial_i\phi) + \frac{\partial A}{\partial(\partial_t\phi)}\delta(\partial_t\phi)$$

$$+\frac{\partial A}{\partial\phi^*}\delta\phi^* + \sum_{i=1}^{3N}\frac{\partial A}{\partial(\partial_i\phi^*)}\delta(\partial_i\phi^*) + \frac{\partial A}{\partial(\partial_t\phi^*)}\delta(\partial_t\phi)$$

plus additional quadratic and higher terms in $\delta\phi$, $\delta\phi^*$ Now the variation "operator" δ and partial differentiation commute so that

$$\delta(\partial_i\phi) = \partial_i(\delta\phi) = \epsilon\partial_i\eta$$

$$\delta(\partial_t\phi) = \partial_t(\delta\phi) = \epsilon\partial_t\eta$$

Thus, to first order in ϵ

$$\delta A = \epsilon\left(\frac{\partial A}{\partial\phi}\eta + \sum_{i=1}^{3N}\frac{\partial A}{\partial(\partial_i\phi)}\partial_i\eta + \frac{\partial A}{\partial(\partial_t\phi)}\partial_t\eta\right.$$

$$\left.+\frac{\partial A}{\partial\phi^*}\eta^* + \sum_{i=1}^{3N}\frac{\partial A}{\partial(\partial_i\phi^*)}\partial_i\eta^* + \frac{\partial A}{\partial(\partial_t\phi^*)}\partial_t\eta\right)$$

and so

$$\delta\mathcal{A} = \epsilon\int\delta A\,dV\,dt$$

to first order in ϵ . Now we are concerned to express the integrand δA in such a way that the arbitrary variational functions η, η^* appear as a *factor* in the integrand in order that a general statement can be made about the rest of the integrand. We must therefore eliminate the terms in $\partial_i\eta$, $\partial_t\eta$ etc. by carrying through part of the integration. Taking a typical one of the terms containing time derivatives of η and integrating by parts:

$$\int\frac{\partial A}{\partial(\partial_t\phi)}\partial_t\eta\,dV\,dt = \left[\frac{\partial A}{\partial(\partial_t\phi)}\eta\right] - \int\partial_t\left(\frac{\partial A}{\partial(\partial_t\phi)}\right)\eta\,dV\,dt$$

where the first (integrated) term contributes to the boundary conditions on ϕ to be discussed later. This satisfies the requirement that η appear as a factor in the integrand

A typical member of the terms involving spatial derivatives $\partial_i\phi$ is:

$$\int\frac{\partial A}{\partial(\partial_i\phi)}\partial_i\eta\,dV\,dt = \int\sqrt{g}\frac{\partial A}{\partial(\partial_i\phi)}\partial_i\eta\,dQ\,dt$$

Again, integrating by parts and noting the additional complication of the presence of $\sqrt{g}$ we obtain

$$\int \sqrt{g} \frac{\partial A}{\partial(\partial_i \phi)} \partial_i \eta dQ dt = \left[\frac{\partial A}{\partial(\partial_i \phi)} \sqrt{g} \ \eta \right]$$
$$- \int \partial_i \left(\sqrt{g} \frac{\partial A}{\partial(\partial_i \phi)} \right) \eta dQ dt$$

Again the first term contributes to the boundaries. Using these typical terms the expression for $\delta \mathcal{A}$ to first order in ϵ is

$$\delta \mathcal{A} = \epsilon \int \eta \left(\frac{\partial A}{\partial \phi} - \sum_{i=1}^{3N} \frac{1}{\sqrt{g}} \partial_i \left(\sqrt{g} \frac{\partial A}{\partial(\partial_i \phi)} \right) - \partial_t \frac{\partial A}{\partial(\partial_t \phi)} \right) dV dt$$

(plus an expression of identical form in η^* and ϕ^*) for arbitrary variations η, η^* in ϕ, ϕ^*. The condition $\delta \mathcal{A} = 0$ and the fact that η, η^* are arbitrary can only be jointly satisfied if the factors multiplying η and η^* in the integrand are identically zero; i.e. vanish for all values of the q^i and t. This gives two equations, one for the multiplier of η and one for the multiplier of η^*. Since ϕ and ϕ^* appear symmetrically in A both equations are of the same form:

$$\frac{\partial A}{\partial \phi} - \sum_{i=1}^{3N} \frac{1}{\sqrt{g}} \partial_i \left(\sqrt{g} \frac{\partial A}{\partial(\partial_i \phi)} \right) - \partial_t \left(\frac{\partial A}{\partial(\partial_t \phi)} \right) = 0$$

These equations are the Euler-Lagrange equations which, together with the boundary terms, fix ϕ and ϕ^*. The equations above fixes ϕ and ϕ^* in the region of space and time over which the integration, defining $\mathcal{A}$ in terms of A, is carried out.

In the case with which we are concerned the integrand A is the difference between the Hamiltonian density and the energy density. Classically in general co-ordinates this is

$$A = H - E = T + V - E = \frac{1}{2m} \sum_{k,l=1}^{3N} g^{kl} p_k p_l + V(q^i) - E$$

where the g^{kl} are the elements of the metric tensor and p_k, p_l are the momentum components conjugate to the general co-ordinates q^k, q^l. The potential energy function V has been written as a function of the spatial co-ordinates only, but the derivation is still valid for those exceptional occasions when V depends on t. In quantum theory we use the Schrödinger form of the complex action to obtain the kinetic energy density and the energy density

$$T = \frac{|\psi|^2}{2m} \sum_{k,l=1}^{3N} g^{kl} \left(\frac{\partial S'^*}{\partial q^k} \right) \left(\frac{\partial S'}{\partial q^l} \right)$$

$$= \frac{\psi^* \psi}{2m} \sum_{k,l=1}^{3N} g^{kl} \left(\frac{-i}{\psi^*} \frac{\partial \psi^*}{\partial q^k} \right) \left(\frac{-i}{\psi} \frac{\partial \psi}{\partial q^l} \right)$$

$$= \frac{1}{2m} \sum_{k,l=1}^{3N} g^{kl} \left(\frac{\partial \psi^*}{\partial q^k} \right) \left(\frac{\partial \psi}{\partial q^l} \right)$$

and

$$E = \psi^* \psi \frac{i}{\psi} \frac{\partial \psi}{\partial t} = i\psi^* \frac{\partial \psi}{\partial t}$$

giving

$$A = \frac{1}{2m} \sum_{k,l=1}^{3N} g^{kl} \left(\frac{\partial \psi^*}{\partial q^k} \right) \left(\frac{\partial \psi}{\partial q^l} \right) + V(q^1, q^2, \ldots) \psi^* \psi - i\psi^* \frac{\partial \psi}{\partial t}$$

The quantities needed to generate an Euler-Lagrange equation for ψ are:

$$\frac{\partial A}{\partial \psi^*} = V\psi - i\frac{\partial \psi}{\partial t}$$

$$\frac{\partial A}{\partial(\partial_j \psi^*)} = \frac{1}{2m} \sum_{l=1}^{3N} g^{jl} \frac{\partial \psi}{\partial q^l}$$

$$\frac{\partial A}{\partial(\partial_t \psi^*)} = 0$$

which, when substituted into the general equation, give an equation determining ψ:

$$\left(V\psi - i\frac{\partial \psi}{\partial t} \right) - \sum_{l=1}^{3N} \left(\frac{1}{\sqrt{g}} \frac{\partial}{\partial q^j} \sqrt{g} \sum_{j=1}^{3N} g^{jl} \frac{\partial \psi}{\partial q^l} \right) = 0$$

which may be re-arranged to

$$-\frac{1}{2m} \sum_{j,l=1}^{3N} \frac{1}{\sqrt{g}} \frac{\partial}{\partial q^j} \left(\sqrt{g} g^{jl} \frac{\partial \psi}{\partial q^l} \right) + V\psi = i\frac{\partial \psi}{\partial t}$$

which is Schrödinger's equation for ψ in general co-ordinates q^i and t. The "equation" for ψ^* (obtained by an analogous procedure) is not sufficient to determine ψ^*; being a trivial degenerate case.

Now the first (spatial derivative) term is recognisable as the full expression for the Laplacian operator ∇^2 in general co-ordinates; and so the equation can be made to look more compact and invariant with respect to co-ordinate systems by writing it as

$$-\frac{1}{2m}\nabla^2 + V\psi = i\frac{\partial\psi}{\partial t}$$

which is its usual form, or

$$-\frac{1}{2m}\nabla^2\psi + V\psi - i\frac{\partial\psi}{\partial t} = 0$$

If, in this geometrical spirit, we re-write the kinetic energy expression in terms of vector operators:

$$T = \frac{1}{2m}|\nabla\psi|^2 = \frac{1}{2m}\sum_{j,l=1}^{3N}\partial_j\psi^* g^{jl}\partial_l\psi$$

hence

$$A = \frac{1}{2m}|\nabla\psi|^2 + \psi^*\psi V - i\psi^*\frac{\partial\psi}{\partial t}$$

we can see that there is a straightforward recipe for generating the Schrödinger equation from the Schrödinger Condition:

$$|\nabla\psi|^2 \rightarrow -\nabla^2\psi$$
$$\psi^*\psi V \rightarrow V\psi$$
$$i\psi^*\frac{\partial\psi}{\partial t} \rightarrow i\frac{\partial\psi}{\partial t}$$

which arises because of the simple form of A as a function of ψ^* and ψ.

Further, if we multiply the Schrödinger equation from the left by ψ^* giving

$$\psi^*\left(-\frac{1}{2m}\nabla^2\psi\right) + V|\psi|^2 = i\psi^*\frac{\partial\psi}{\partial t}$$

the similarity to the original A is even more striking: only the kinetic energy expression is replaced by another term. This similarity has proved so striking as to have been extremely misleading, historically speaking. By concentrating attempts to obtain a physical interpretation on the Schrödinger *equation* instead of the more fundamental Schrödinger variational *condition*, paradoxes have crept into the physical interpretation of Schrödinger's quantum mechanics which are avoidable.

Unfortunately, the very appearance of the Schrödinger equation, particularly in the form given above, has the appearance of an "energy balance"equation which has led to the term

$$\psi^* \left(-\frac{1}{2}\nabla^2 \right) \psi$$

being interpreted as the kinetic energy density which it is not. The kinetic energy density is, of course,

$$T = \frac{1}{2m}|\nabla\psi|^2$$

which, for example, is always positive as kinetic energy must be; whereas the expression involving the Laplacian may be either positive or negative. This confusion is compounded by the fact that for isolated bound particles the two expressions above integrate to the same mean value!

$$\int \psi^* \left(-\frac{1}{2m}\nabla^2 \right) \psi dV = \int \frac{1}{2m}|\nabla\psi|^2 dV$$

because, for such systems, the difference between the two densities integrates to zero.

Taking such a case — a constant energy system — the Schrödinger equation becomes

$$-\frac{1}{2m}\nabla^2\psi + V\psi = E\psi$$

from which

$$\psi^* \left(-\frac{1}{2m}\nabla^2\psi \right) + V|\psi|^2 = E|\psi|^2$$

Now, the argument goes, since the energy is constant the sum of the Kinetic and potential energies must be constant and this only obtains if the kinetic energy is given by the Laplacian expression so that

$$T + V = E = \text{ constant}$$

in the above expression. If the expression

$$T = \frac{1}{2m}|\nabla\psi|^2$$

is used, then the sum of the kinetic energy density and the potential energy density is not constant which contradicts the original assumption.

But the Schrödinger equation arises from the Schrödinger Condition which requires that the Hamiltonian density (the sum of the kinetic and potential energy densities) be equal to the energy density *only on the average* not point-by-point in space (and time). As we have seen above, both "candidates"for the kinetic energy density satisfy this condition so there is no difficulty. To insist that the sum of the kinetic and potential energies be equal to the total energy at all points in space is to contradict the central Schrödinger Condition that only the *means* of these quantities are required to be equal in quantum mechanics.

> In summary the Schrödinger equation is an equation which determines the function ψ and the physical interpretation of the dynamical quantities of the theory are fixed in advance by the interpretation of the more fundamental Schrödinger Condition. In setting up the Schrödinger Condition, the particle probability density, the kinetic energy density, the potential energy density and the total energy density are all defined independently in terms of the complex action S' or ψ and the Schrödinger Condition *fixes relationships among the average of these densities.*

The Schrödinger equation is a kind of "auxiliary"equation which determines ψ and can only carry the physical interpretation given in the fundamental variational integral. If this is forgotten, by fixing attention on the Schrödinger equation only, then paradoxes result — like negative kinetic energy. It is not a paradox that in an isolated (constant energy) system the sum of the kinetic and potential energies is not equal to the total energy at all points in space because it is *precisely* this freedom of the total energy density to deviate from the Hamiltonian density which distinguishes Schrödinger's quantum mechanics from classical mechanics. It is not a paradox but it is hard to understand - in short it is a call further creative work!

It is increasingly obvious that questions of the physical interpretation of the quantities arising in the manipulation and solution of the Schrödinger equation are coming to the fore, particularly those problems associated with momentum densities and with probabilities in general. The next chapter tries to address some of the major difficulties.

The one outstanding matter to be dealt with is the interpretation of the fixed integrals which have, so far, been called simply "boundary terms". If the real and imaginary parts of the function S' do indeed carry the interpretation that we have put on them as far as momentum components are concerned then there are some elementary consequences. For, if the momentum density is expressed as

$$\psi^* \left(-ik\frac{\partial \psi}{\partial q^j} \right) = d_p \quad \text{(say)}$$

$$= Re(d_p) + iIm(d_p)$$

then on integration over all space we should obtain the mean momentum of the ensemble (assuming normalised ψ), for that particular momentum component. But we have said that this mean value is carried by $Re(S')$ so that, taking account of the explicit appearance of i in the left-hand-side above, the mean value of the momentum component should be given by the real part of the above expression. That is the imaginary part should *integrate to zero* carrying, as we have asserted, only the deviations of the momentum component from the mean. Of course, since we expect the mean value of any momentum component to be a real number it would be embarassing for the mean to turn out to be a complex number quite independently of the physical interpretation which we have given to the two components of the momentum distribution. In fact, as we shall see, this always proves to be the case. This reinforces the interpretation of the two components of the momentum distribution: the real part of the distribution (arising from the real part of S') is the distribution of the ensemble as a whole which, when integrated generates the mean value, while the imaginary part of the distribution gives the distribution of deviations from the momentum of the ensemble as a whole which, when integrated gives zero. Insofar as dynamical variables are functions of co-ordinates and momenta, that is in Schrödinger's theory spatial distributions of particles and their momentum distributions, we expect similar considerations to apply to other dynamical variables. In particular, recall that the energy function itself may be regarded as conjugate to the time variable and so, when the dynamical law is imposed and the Hamiltonian is equal to the energy (on average) the real and imaginary parts of the energy distribution should have a similar interpretation to the above for momentum components. In practice, much of quantum dynamics is concerned with isolated conservative systems for which each member of the ensemble has the same energy so that the resulting function S' has no imaginary part depending on t and, of course, no deviations from the mean energy.

We can easily see that the interpretation of the real and imaginary parts of S' is at least mathematically consistent by investigating the condition that a given ensemble represented by S' has a constant momentum component per particle. The condition is simply

$$-\frac{ik}{\psi}\frac{\partial \psi}{\partial q^j} = \lambda \quad \text{(say)}$$

That is

$$\frac{\partial \psi}{\partial q^j} = i\lambda\psi$$

(absorbing k into the constant λ) which has the simple solution

$$\psi = C \exp(i\lambda q^j) = A \exp(iS) \quad \text{(say)}$$

Where C is chosen to normalize ψ. That is the action (S') is wholly real for an ensemble which is homogenous in one momentum component. Similarly for an ensemble of members all having the same energy:

$$\frac{ik}{\psi}\frac{\partial \psi}{\partial t} = \quad constant$$

$$\frac{\partial \psi}{\partial t} = iE\psi$$

$$\psi = C \exp(iEt)$$

for any value of the constant energy.

These results are independent of any dynamical laws, they are merely the result of Schrödinger's generalisation of S to S' together with the interpretation of the momentum as the momentum per particle of an ensemble. The equation for an energy-homogeneous ensemble above becomes of interest when the dynamical law is imposed and the constant energy ensemble is related to the distribution of the Hamiltonian function. There are no corresponding equations for the momentum components so that the form of the momentum-homogeneous distributions remain simply identities.

Chapter 4

Interpretation

4.1 Introduction

In the last chapter the transition has been made from classical to quantum mechanics by the requirement that, in Schrödinger's dynamics, the Hamilton-Jacobi equation must be satisfied only on the average in space and time. We have also embarked on an interpretation of the various quantities appearing in Schrödinger's theory albeit on an informal and intuitive basis. There are now several possible ways forward:

1. An investigation of the formal properties of the equations and identities of the Schrödinger theory with a view to extracting the "structure"of the theory for a more rigorous and, perhaps, more general formulation.

2. A continued investigation of the interpretation of the theory and a more precise definition of the referents of the theory and a more careful statement of the results of the last chapter.

3. Application of the Schrödinger theory to particular systems.

In fact, we shall not be concerned at all with (3) except insofar as it bears on (1) and (2); the possible applications are very numerous and extremely successful and it was these very successes which guaranteed the acceptance of Schrödinger's theory long before an adequate interpretation of the formalism was available.

This chapter will be concerned with (1) and (2) and, in particular with the connection between (1) and (2): how the interpretation of the theory bears on the validity and utility of any formal structures which may be abstracted from that theory. What are we to make, for example, of the fact that the Schrödinger equation has, no doubt, infinitely many solutions? Is there a relationship between the families of solutions of the Hamilton-Jacobi equation obtained by separation methods and the "corresponding"solutions of Schrödinger's equation? How do we interpret those amongst the solutions which cannot be normalised to unity? Can we distinguish between equations and identities in the theory as we did in classical mechanics? What concepts replace the familiar concepts of a particle's position, energy, momentum and (as yet unmentioned) velocity in classical mechanics? These and many other pressing scientific problems are not touched by the ability of the Schrödinger equation to yield numerical results in essentially perfect agreement with experiment.

Superposed on these problems which are particular to Schrödinger's theory is the all-pervasive problem which arises whenever an attempt is made to express a scientific theory in terms of a known mathematical structure: does the formal "generalisation"reveal the real structure of the theory without the inclusion of sources of scientific difficulty?

In our reflections on classical mechanics and on the transition to Schrödinger's mechanics it has not proved necessary to pause to consider the nature of the mathematical structures used or their possible impact on the science involved since - along with Newton, Lagrange, Hamilton, Jacobi and Schrödinger - we have simply *assumed* the validity and utility of the differential and integral calculus. In a word it has been silently assumed that the methods and results of classical analysis *model* the continuity and structure of real space and time. Moreover, it was assumed that this modelling is done in the usual sense of abstraction; even if the use of the real numbers (for example) to model each dimension of real space *excludes* some of the properties of real space, the opposite is not true - the use of the reals to model a line does not surreptitiously *include* any properties which are not possessed by physical space. This may be incorrect, of course, but it is the classical viewpoint.

The possibilities of unintentionally including extraneous material into a scientific theory is much more real when the more powerful techniques of mathematics are used. The equations of classical mechanics specify *local* conditions among the mechanical quantities: they are differential equations. If it proves possible to generate these equations from a *global condition* — a variation principle, for example — then there is always the possibility, even likelihood, that the global condition will generate more differential equations and boundary conditions than the one which was its ostensible "source".

This latter trap is only a danger in classical mechanics where we start with $F = ma$ and generalise the theory by the admission of more and more classes of "co-ordinate". Perhaps the most obvious place where we depend on a global condition is in the development of the transformation theory where it is required that the generating function be capable of being developed as a function of time only which is the direct result of the variational formulation of mechanics.

In Schrödinger's mechanics, by contrast, the *starting point* is a global condition: the equality of the mean values of the Hamiltonian density and the energy density. The Schrödinger *equation* which is the (local) Euler-Lagrange equation of this global condition (together with some boundary conditions) and is therefore, at a lower level than the fundamental Schrödinger condition. But this condition on the mean values of the Hamiltonian and energy densities is based on the preceeding classical mechanics. If Schrödinger's theory is considered to depend on classical *mechanics* then any artifacts in classical mechanics may be carried forward. If, however, Schrödinger's mechanics in viewed as independent of classical mechanics, depending only on historically-established classical mechanical *concepts* (not physical laws) then it is an independent global theory from the outset.

In the context of considering the validity of abstraction and generalisation it is perhaps not out of place to be prepared for the rather eccentric use to which mathematicians put the word "representation". In ordinary English, if A is a "representation"of B then B is richer and more complex then A; it contains more structure than A. For example, a wiring diagram is a "representation"of the electrical system in a car or building; a flow chart is a "representation"of an industrial process; a portrait is a "representation"of a child. In all these cases the "representer"is an abstraction of some properties of the "represented"and is used to show only one aspect of a more complex entity under study. It might therefore be reasonably expected that, in mathematics, the statement "A is a representation of B"might be a paraphrase of something like "A is abstracted from B"as it is in both everyday and scientific usage.

In fact the mathematical usage is exactly the opposite; the abstraction is taken to be the more basic or fundamental entity (the represented) and the concrete entity as the representer. Thus, line elements in real space are a "representation"of an abstract vector space; rotations of solid bodies in real space are a "representation"of group theory and (to look ahead) the solutions of the Schrödinger equation are a "representation"of Hilbert space. One assumes that this inverted way of expressing an idea is to make allowance for the undoubted fact that the same mathematical structure may be abstracted from many (perhaps unrelated) real structures and processes and not in the (Platonic) belief that the mathematical structures *are* more fundamental.

This terminology will not cause any difficulty in actual applications of mathematics to scientific theories but it can give an unfortunate philosophical slant to the interpretation of those theories. If a mathematical or logical structure which is contained in (can be abstracted from) a scientific theory is regarded as more fundamental than that scientific theory then the physical interpretation of the quantities in the theory becomes difficult and may come to be regarded as arbitrary and this view can be reinforced by the generation of intuitive paradoxes in the abstract theory which were not present in the original, physically interpreted, theory.

The use of the extremely powerful methods of mathematics has more than just a manipulational value in physical science; they are a guide to intuition and an aid to concept formation. But mathematics is neither the source nor the destination of scientific theories. In spite of eminent opinion to the contrary, therefore, it is not possible for there to be any "Mathematical Foundations of Quantum Mechanics". If the architectural analogy is to be used at all for the role of mathematics in the sciences it is more akin to the wiring, plumbing and heating — all those services which make a building convenient and pleasant to live in — than to the foundations. A more realistic metaphor is between mathematics and the tools with which the edifice is constructed: the building's ultimate shape is determined just as much by the techniques available as it is by the use for which it is intended.

In the next short section some of these points are exemplified by reference to material met earlier and in later sections we will attempt to address the more general problem of abstractions from the Schrödinger theory.

4.2 More on Kinetic Energy and Angular Momentum

Every Schrödinger equation, whatever the potential in which the particles move, contains a Laplacian operator ∇^2 which arises from the presence of the kinetic energy density $|\nabla\psi|^2$ in the variational problem. In the last chapter it was stressed that the Laplacian operator is not a "kinetic energy operator" i.e. the kinetic energy density of a particle is not

$$\psi^* \left(-\frac{1}{2}\nabla^2 \right) \psi \quad \text{but} \quad \frac{1}{2}|\nabla\psi|^2$$

however the Laplacian is always *associated* with the kinetic energy of each particle as that operator in the Schrödinger equation which is the local operator replacing the global condition on the kinetic energy density in the variational problem.

The explicit relationship between the "Laplacian density" and the kinetic energy density is illuminating since it highlights the way in which the global Schrödinger condition bears on the boundary conditions of the local equation and on the associated physical interpretation. For a single particle of unit mass we have

$$\frac{1}{2}|\nabla\psi|^2 = \frac{1}{2}\left[\psi^*\left(-\frac{1}{2}\nabla^2\right)\psi + \psi\left(-\frac{1}{2}\nabla^2\right)\psi^*\right] + \frac{1}{4}\nabla^2|\psi|^2$$

If the two expressions in braces on the right-hand-side were equal then a direct relationship could be obtained. This equation cries out for the application of Green's theorem; integrating both sides we obtain an expression for the average value of the kinetic energy in terms of integrals involving the Laplacian:

$$\int\frac{1}{2}|\nabla\psi|^2 dV = \frac{1}{2}\left[\int\psi^*\left(-\frac{1}{2}\nabla^2\right)\psi dV + \int\psi\left(-\frac{1}{2}\nabla^2\right)\psi^* dV\right]$$
$$+\frac{1}{4}\int\nabla^2|\psi|^2 dV$$

Now the two expressions in the braces may each be integrated by parts (twice) to show that they are equal if ψ and $\nabla\psi$ vanish on the boundary of the region in which the particle moves or that "incoming" and "outgoing" components of ψ and $\nabla\psi$ cancel. Under these conditions the Divergence Theorem guarantees that the remaining term, involving the Laplacian of $|\psi|^2$, is zero. That is, for bound states or for systems in which there is no net creation of particle density the equality

$$\int\frac{1}{2}|\nabla\psi|^2 dV = \int\psi^*\left(-\frac{1}{2}\nabla^2\right)\psi dV$$

holds. The *mean value* of the kinetic energy may be obtained by integrating the Laplacian density

$$\psi^*\left(-\frac{1}{2}\nabla^2\right)\psi$$

or by integrating the kinetic energy density.

But the equality of the integrals does not imply the equality of the integrands - far from it, as $|\nabla\psi|^2$ is always positive while the Laplacian density may be negative. The mean value of the kinetic energy is always

$$\int\frac{1}{2}|\nabla\psi|^2 dV$$

and usually

$$\int\psi^*\left(-\frac{1}{2}\nabla^2\right)\psi dV$$

but the kinetic energy density is always

$$\frac{1}{2}|\nabla\psi|^2$$

and never

$$\psi^*\left(-\frac{1}{2}\nabla^2\right)\psi$$

As we noted earlier this removes the apparent paradox of negative kinetic energy.

To emphasise these points it is useful to look at equations defining ensembles of constant momentum, of constant kinetic energy and the Schrödinger equation a free particle:

1. The momentum per particle in Schrödinger's theory is

$$\frac{-i}{\psi}\nabla\psi$$

and so, in one dimension, an ensemble of constant momentum per particle is a solution of

$$\frac{-i}{\psi_1}\frac{d\psi_1}{dx} = \lambda \quad \text{(say)}$$

i.e.

$$\frac{d\psi_1}{dx} = i\lambda\psi_1$$

2. The kinetic energy density is

$$\frac{1}{2}|\nabla\psi|^2$$

and so the kinetic per particle is

$$\frac{1}{|\psi|^2}\frac{1}{2}|\nabla\psi|^2$$

and, again in one dimension, an ensemble of constant kinetic energy per particle is a solution of

$$\frac{1}{|\psi_2|^2}\frac{1}{2}\left|\frac{d\psi_2}{dx}\right|^2 = \mu \quad \text{(say)}$$

i.e.

$$\left|\frac{d\psi_2}{dx}\right|^2 = 2\mu|\psi_2|^2$$

3. The Schrödinger equation for a free particle is

$$-\frac{1}{2}\nabla^2\psi = i\frac{\partial\psi}{\partial t}$$

which separates to give a one-dimensional equation like

$$-\frac{1}{2}\frac{d^2\psi_3}{dx^2} = \nu\psi_3 \quad \text{(say)}$$

where ψ depends on x only

The solutions of the first and last of these equations are simple:

- $\psi_1 = A\exp(i\lambda x)$

- $\psi_3 = B\cos(\nu x) + C\sin(\nu x)$

Where $|\psi_1|^2$ is constant and $|\psi_3|^2$ is not constant. Now ψ_1 solves equation (2) with $\mu = \lambda^2$ but ψ_3 does not solve (2). The physical reason for this is clear enough; an ensemble of constant momentum per particle is necessarily an ensemble of constant kinetic energy per particle. The converse is equally obviously false; an ensemble of constant kinetic energy per particle may be composed of members with the same magnitude of angular momentum but in either of the two directions (or in any direction in three-dimensional space).

It is worth remarking that of the three expressions (1) - (3) above, the first two are *identities* being merely the definitions of certain ensembles independent of any dynamical law. The Schrödinger equation for a free particle is, however, just that; *an equation* for the energy and distribution of a free particle - albeit a rather trivial application of the dynamical law.

This interpretation of the total kinetic energy also holds for the individual "components" of the kinetic energy expression when this (scalar) expression is evaluated in some particular co-ordinate frame. In particular, the analysis of the kinetic energy for centrally symmetric systems in spherical polar or sphero-conal co-ordinates yields an operator component of the Laplacian corresponding to the angular kinetic energy. The study of angular kinetic energy has traditionally being carried out under the "Angular Momentum" heading since the square of the total angular momentum and the angular kinetic energy are identical except for an angle-independent factor: the moment of inertia mr^2 for a particle of mass m distance r from the origin. However, the square of the total angular momentum "vector" is not an angular momentum and, *a fortiori*, is not a canonical angular momentum component. To emphasise the non-canonical nature further, note that the relationship between the kinetic energy and canonical momenta is:

$$T = \sum_{k,l=1}^{3N} g^{kl} p_k p_l$$

an expression quadratic in the canonical momenta, showing conclusively that ℓ^2 can never be considered as an angular momentum -it is the angular kinetic energy.

Both of the examples in this section: the question of the kinetic energy "operator" and the role of the angular kinetic energy arise from an *uncritical interpretation of suggestive formalisms.*

On the one hand, since the momentum distribution is given by

(particle distribution) × *(momentum densities per particle)*

$$\psi^* \psi \frac{1}{\psi}(-i\nabla\psi) = \psi^*(-i\nabla)\psi$$

the analogy with

$$\psi^*(-\frac{1}{2}\nabla^2)\psi$$

as the kinetic energy density has sometimes proved too tempting to resist, particularly if one is concerned *only* with mean values and not with actual densities.

On the other hand, the vanishing of the Poisson bracket

$$[\ell^2, \ell_\alpha]$$

suggests a canonical role for ℓ^2 if taken in isolation from the physical interpretation *and* the other two Poisson brackets.

It is with these cautionary tales in mind that it is proposed to approach the formalisation of the Schrödinger theory. There are two possible avenues to pursue:

1. An abstraction of the formal properties of the Schrödinger *equation* and its solutions

2. Formalisation of the properties of the various densities which are used in the Schrödinger Condition: momentum density, kinetic energy density, probability density etc.

Naturally these two investigations must ultimately come together but each will benefit from some initial independent investigation.

4.3 Operators or densities?

It has been shown earlier and stressed on a number of occasions that the Laplacian operator cannot be considered to be the "kinetic energy operator" except in one very specialised sense. The "Laplacian density"

$$\psi^* \left(-\frac{1}{2}\nabla^2 \right) \psi$$

does not model the kinetic energy density for a single particle but the mean value of the kinetic energy is given by

$$\int \psi^* \left(-\frac{1}{2}\nabla^2 \right) \psi dV$$

when the integration is carried out over all space and when ψ obeys the most usual boundary conditions. This has the corollary, or course, that the so-called Hamiltonian operator

$$\hat{H} = -\frac{1}{2}\nabla^2 + V$$

(for a single particle) may only be considered to be the "energy operator" under even more restrictive conditions:

1. The true energy operator is $i\partial/\partial t$ and the equality of the Hamiltonian density and the energy density in the mean is the global Schrödinger Condition which generates the local Schrödinger equation.

2. The Hamiltonian operator abstracted from the Schrödinger equation even under the condition (1) is only the energy operator in the same sense that the Laplacian is the kinetic energy operator : its mean value is the correct mean value of the energy but not the correct energy density.

This section is devoted to a brief examination of the possibility of replacing the probability densities for dynamical properties of a (single-particle for simplicity) system in Schrödinger's mechanics by certain operators. That is, we attempt to decide if the "dynamical variables" of a quantum system may be modelled by operators rather than functions.

In Schrödinger's theory the fundamental quantities which replace the classical dynamical variables like position, momentum, potential energy, kinetic energy etc. are probability distributions or, more colloquially, densities. These densities arise in the following way. For a one-particle system (an ensemble of single particles in well-defined environments) the probability that the particle be in a particular region W of ordinary 3-space is

$$\int_W \psi^* \psi dV$$

if ψ is normalised to unity so that

$$\int \psi^* \psi dV = 1$$

Now, if the probability distribution in space of some quantity A is required and the function ψ is known there are just two possibilities:

1. A is a function of the co-ordinates (and possibly t) but not of the momenta of the particle

2. A is a function of both the co-ordinates and the momenta (and possibly t) or just a function of the momenta.

If (1) obtains then it only requires the applications of classical probability theory to define

$$\int_W A\psi^* \psi dV$$

as the "amount of A" in region W and so to use $A\psi^* \psi$ as the spatial distribution of the property A, which may, with impunity, be written

$$A\psi^* \psi = \psi^* A\psi = \psi^* \psi A$$

since only multiplication of A by the particle density is involved. In such cases the distribution of A *per particle* is just

$$\frac{\psi^* A\psi}{\psi^* \psi} = A$$

that is the distribution of A only involves ψ in the *passive* sense of merely reflecting the distribution of the particle. Indeed quantities which behave in this way may be thought of as properties of the particle's environment rather than of the particle. The most common example in actual applications is when A is the potential energy function of a conservative system: $A = V(q^k)$. In this case the very name potential energy suggests that $V(q^k)$ is independent of the particle density. In a word these quantities are not operators but mere multiplying factors which are *weighted* in the theory by the particle probability densities.

However, if the quantity A involves momenta (perhaps in addition to any dependence on co-ordinates) then the situation is more complex if any structure is to be abstracted from the resulting densities. The way to construct a density in these cases is obvious enough. Let A be a function of the q^i and p_i ; expressed in classical mechanics as $A(q^i, p_i)$. Then in Schrödinger's theory the momentum per particle is

$$-\frac{i}{\psi}\frac{\partial\psi}{\partial q^k}$$

so that the density per particle of A is just

$$A(q^k, -\frac{i}{\psi}\frac{\partial\psi}{\partial q^k})$$

and the overall density of A is this expression multiplied by the actual particle density as before

$$\psi^*\psi A(q^k, -\frac{i}{\psi}\frac{\partial\psi}{\partial q^k})$$

This expression has exactly the same structure as the previous momentum-free expression with the important exception that the density of A *per particle* depends on ψ: i.e. is a property of the particle not just its environment. This is not surprising since ψ contains both the real and imaginary action : the particle distribution and its momentum distribution. Of course when A is H (the classical Hamiltonian function), this procedure is precisely the one we used to set up the dynamical Law in Schrödinger's mechanics, so any other dynamical variable densities should be compatible with the Hamiltonian density.

The real difficulties arise when one compares the form of quantities *linear* in momenta with those which are *non-linear* in such momenta. If A is linear in the canonical momentum per particle

$$-\frac{i}{\psi}\frac{\partial\psi}{\partial q^k}$$

i.e. if

$$A(q^k, \alpha p_k) = \alpha A(q^k, p_k)$$

then

$$A(q^k, -\frac{i}{\psi}\frac{\partial\psi}{\partial q^k}) = -\frac{i}{\psi}A(q^k, \frac{\partial\psi}{\partial q^k})$$

$$= -\frac{i}{\psi}\left[a(q^k)\frac{\partial\psi}{\partial q^k} + b(q^k)\right] \quad \text{(say)}$$

and the total density is

$$\psi^*\psi a(q^k, -\frac{i}{\psi}\frac{\partial \psi}{\partial q^k}) = \psi^*\left[-i(a(q^k)\frac{\partial}{\partial q^k} + b(q^k))\right]\psi$$

$$= \psi^*\hat{A}\psi \text{ (say)}$$

with

$$\hat{A} = -i\left[a(q^k)\frac{\partial}{\partial q^k} + b(q^k)\right]$$

which is an expression reminiscent of the momentum density

$$\psi^*\left(-i\frac{\partial}{\partial q^k}\right)\psi = \psi^*\hat{p}_k\psi \quad \text{(say)}$$

Thus, when dynamical variables are linear in momentum components then the density of such variables may be reduced to a form which is the same as that of the momentum density. But this is not all surprising or novel since variables linear in momenta, formally speaking, are scarcely different from momenta.

In contrast, any quadratic or higher dependence of A on momentum components precludes the reduction of the A-density to the form

$$\psi^*\hat{A}\psi$$

where $\hat{A}$ is a linear differential operator. At its simplest, this is merely because

$$\left(\frac{\partial \psi}{\partial q^k}\right)^2$$

cannot be reduced to a form

$$f(q^k)\hat{T}\psi$$

where $\hat{T}$ is a linear differential operator. It may happen, as it were by co-incidence, that such a reduction is possible in a special case but it is quite clear that a reduction in general is not possible.

As will be seen shortly, the formal properties of linear differential operators are so attractive that some considerable effort has been spent over the years into attempting to reconcile the linear differential momentum operator $-i\nabla$ with linear differential operators which will model A-densities in the sense that $\psi^*\hat{A}\psi$ models the A-density in the same way that $\psi^*(-i\nabla)\psi$ models momentum-density. Since, as we have seen, this is impossible in general for the momentum density per particle of the Hamilton-Jacobi-Schrödinger type

$$-\frac{i}{\psi}\frac{\partial\psi}{\partial q^k}$$

attempts have been made to use new momentum operators (i.e. different expressions for the momentum density) in order that these operators may generate A-densities modelled by $\psi^*\hat{A}\psi$ for linear $\hat{A}$.

It must be said quite frankly that these efforts, which have met with some limited success in particular co-ordinate systems, are vain and are vain on two counts: [1]

1. Without the introduction of some *ad hoc* hypothesis in the construction of the operator $\hat{A}$ none of the attempts (which must retain the *differential* form of the momentum density) have succeeded or can succeed in general.

2. The main motivation for these efforts has always been the desire to construct a linear kinetic energy operator which is quadratic in the components of the canonical momenta i.e. which models the classical relationship

$$T = \sum_{k,l=1}^{3N} p_k p_l g^{kl}$$

That is, a way of generating the (scalar) Laplacian operator from the canonical momentum "operators"

The technical difficulties which arise are due to the dependence of the g_{kl} on the co-ordinates q^k which requires the resolution of the problem of the differentiation of a product.

But the Laplacian is not the operator which generates the kinetic energy density! Thus all these efforts, even if successful, would not generate a physically meaningful theory - the mean value of the kinetic energy *only* is given by the integral of the Laplacian and even that under specific boundary conditions on ψ. Even in Cartesian co-ordinates, where the g_{kl} are particularly simple, and so the squares of the canonical momentum components do indeed generate the Laplacian, the success at generating the Laplacian density must be tempered by the knowledge that this is not the kinetic energy density! That is, only in Cartesian co-ordinates is "∇-squared" actually

---

[1] This opinion was clearly stated by Kemble ("Fundamental Principles of Quantum Mechanics"McGraw-Hill) as early as 1937 "It must be frankly admitted at the outset that an examination of the efforts which have been made to set up a general theory answering the above and related questions suggests at times that the possibility that the game is not worth the candle."

the square of ∇; but this has no relevance to the fact that the kinetic energy density is the square of the momentum density and the kinetic energy operator is not the square of the momentum operators.

There is then no possibility of obtaining linear differential operators $\hat{A}$ which depend on the canonical momenta (however reasonably expressed) in the same way as the classical analogue *and* generate the correct A-density by

$$\psi^* \hat{A} \psi$$

If the classical expression for A is $A(q^k, p_k)$, then the quantum expression (in all co-ordinate systems) is

$$\psi^* \psi A(q^k, -\frac{i}{\psi} \frac{\partial \psi}{\partial q^k})$$

in which no ambiguity is involved since only functions occur in the expression.

> Therefore, notwithstanding overwhelming opinion to the contrary, Schrödinger's mechanics cannot be cast into a form in which the dynamical variables are represented by linear differential operators in the sense that the A-density is not $\psi^* \hat{A} \psi$ in general.

Strictly speaking one should make the exception that operators linear in the canonical momenta can be expressed as linear operators but this is a special case of little physical interest and, more important, the generation of functions ψ which are momentum-homogeneous does not contain the *dynamical law*. Such expressions are identities not equations.

This conclusion does not, of course, have any mathematical consequences for the manipulations of the Schrödinger equation. The "Hamiltonian" operator is a linear differential operator and the mathematical consequences of this will be sketched in the next section - our conclusion simply means that the conventional "Hamiltonian density" $\psi^* \hat{H} \psi$ is not the energy density even when ψ solves the dynamical law.

We have been at pains to stress and re-stress the role of the Laplacian and of the Hamiltonian in determining the *mean* kinetic energy and energy respectively. One might therefore take the view that only these average values are important in the theory since, generally speaking, these are the quantities which are accessible to experiment. Thus, the distinctions which have been stressed are so much hair-splitting and any operator $\hat{A}$ which generates the *mean value* of the variable A correctly as

$$\int \psi^* \hat{A} \psi dV$$

might pass muster as "modelling A" and the density is irrelevant. In this view, for example, the Laplacian and the Hamiltonian are perfectly acceptable as kinetic energy and energy operators when ψ solves the Schrödinger equation. What is more we can, by rule of thumb, discover the correct form for the Hamiltonian in Cartesian co-ordinates (not easy in the presence of a vector potential) and so transform to any suitable co-ordinate system at will. There are several possible objections to this empirical point of view:

- It violates the methodological, historical and logical continuity of Schrödinger's theory with the classical theories of Newton, Lagrange, Hamilton and Jacobi.

- Information is thrown away (the densities) and nothing is gained: the manipulational and technical aspects of Schrödinger's equation are untouched and the interpretation of the solutions is muddied with paradoxes like negative kinetic energy.

- It does not work.

The last of these objection may seem a little surprising since the whole point of this empirical interpretation of the Schrödinger theory was that it should work and work, what is more, independently of the Schrödinger Condition, enabling the theory to *start with* the Schrödinger equation and manipulations with operators.

The prescription for the generation of a linear operator which gives the correct average value for the kinetic energy or total energy (of a conservative system) — both of which are quadratic in canonical momenta — comes, not from a definition of the relationship of momentum components to kinetic energy but from the central dynamical law of Schrödinger mechanics. It is the equality of the mean values of the Hamiltonian density and the energy density — a global condition — which generates the local Laplacian operator having the correct mean value (usually; subject to boundary conditions). Now there is no such equation for dynamical variables in general; for an arbitrary

$$A(q^k, -\frac{i}{\psi}\frac{\partial \psi}{\partial q^k})$$

there is no analogue of the Hamilton-Jacobi equation and so no Schrödinger Condition and thus no analogue of the Schrödinger equation. Without this analogue we are at a loss to know how to replace the true A-density

$$\psi^* \psi A(q^k, -\frac{i}{\psi}\frac{\partial \psi}{\partial q^k})$$

by a density with the same mean value of the required form:

$$\psi^* \hat{A} \psi$$

We are at a loss to know if there is one such or many such and certainly at a loss to know the physical interpretation should one be found.

The formal analogy between the identity which determines ψ for an ensemble with constant momentum per particle

$$-\frac{i}{\psi} \nabla \psi = \lambda \quad : \quad \nabla \psi = i\lambda\psi$$

and the Schrödinger equation for a conservative system:

$$\hat{H}\psi = E\psi$$

masks the very fact that the first is an identity and the second is part of Schrödinger's dynamical law determining those ensembles for which the Hamiltonian density has the mean value E; which also happens to be the mean value of

$$\psi^* \hat{H} \psi$$

The properties of the solutions of Schrödinger's equation for conservative systems are all-important in applications of the theory and have dominated the physical interpretation of the theory; the next section sketches some of these properties.

4.4   The Time-Independent Schrödinger Equation

The vast majority and most important applications of the Schrödinger equation are those for which the solutions satisfy the boundary conditions ensuring that the mean value of the kinetic energy is (half of) the mean value of the Laplacian density. A large class of such systems are those whose Hamiltonian density, and therefore Hamiltonian operator, are independent of time: isolated conservative systems, in the main. Such systems have, of course, constant energy and, just as in the case of the Hamilton-Jacobi equation, the Schrödinger equation may be cast into a particularly simple form. If the energy is constant then an ensemble of constant energy per particle has a very simple time dependence since

$$\frac{i}{\Psi} \frac{\partial \Psi}{\partial t} = E \text{ (say)}$$

$$\Psi = \psi(q^k)\exp(-iEt)$$

where $\psi(q^k)$ is independent of time. In this section we wish to make an explicit distinction between the full (time-dependent) solution of the Schrödinger equation which we write Ψ and the time-independent factor written ψ. And, of course,

$$\Psi^*\Psi = \psi^*\psi$$

there being no dependence of the particle distribution on time.

Substituting this expression into the Schrödinger equation - a product of a function of space only and an exponential time-dependent factor - effects a separation of variables leaving a "space-only" partial differential equation known as the time-independent Schrödinger equation. For a single particle this is the familiar

$$\left[-\frac{1}{2m}\nabla^2 - V(q^k)\right]\psi(q^k) = E\psi(q^k) \tag{4.4.1}$$

an eigenvalue problem, since E is a constant by supposition. The solution of the Schrödinger equation is then

$$\Psi(q^k,t) = \psi(q^k)\exp(-iEt) \tag{4.4.2}$$

Now the possibility of admitting electro-dynamic interactions has not yet been investigated and such an investigation will show that terms of a new *form* will have to be included in the Schrödinger equation involving the unit imaginary. But for the moment we may concentrate on 4.4.1 and note that it is entirely a *real* equation and the boundary conditions for systems of most physical interest are real —the vanishing of $|\psi|^2$ at the boundaries of the region for the most part — and so the space-dependent factor $\psi(q^k)$ is a *real function*. The only possible exception to this is the case of solutions of 4.4.1 with the same E which, for the reasons outlined above, will either be real or be complex conjugate pairs which, because of the *linearity* of 4.4.1, means that real solutions may always be found.

This simple fact has some obvious consequences. We have seen earlier that the physical interpretation of the solutions of Schrödinger's equation involves a separation of the "mean"of and "deviations"from the canonical momenta. The means of the canonical momenta conjugate to any co-ordinate is carried by the real part of S' and the deviations from that mean by the imaginary part. Now since energy may be considered to be conjugate to time we note that, since there is no imaginary time-dependence of S' then there are no deviations from the mean-value of E. This is, of course, just a comforting tautology since we have assumed that E is constant; but it does show a consistency with our earlier interpretations. In contrast, there is no imaginary spatial-co-ordinate dependence of S', meaning that the means of all the canonical momenta are zero for ensembles described by the time-independent Schrödinger equation. In the special case of complex conjugate pairs of spatial factors with the same E this simply means that the zero-mean ensemble may be decomposed into momentum-homogeneous ensembles with equal and opposite mean values of canonical momenta. In general, in the absence of complex-conjugate pairs of solutions, the ensembles described by the solutions 4.4.1 have mean energy E and all mean canonical momenta zero.

This is, perhaps a convenient place to investigate the deviations from the mean energy of ensembles described by the time-independent Schrödinger equation. The Schrödinger equation is generated by the equality of the energy density and the Hamiltonian density in the mean. Now the functions 4.4.1 have energy E with no deviations. But the Hamiltonian density does deviate from the mean; since the Hamiltonian density is

$$\frac{1}{2}|\nabla\psi|^2 + |\psi|^2 V(q^k)$$

we have, for the deviations from the mean value of the Hamiltonian density (E, of course)

$$\frac{1}{2}|\nabla\psi|^2 + |\psi|^2 V(q^k) - E|\psi|^2$$

$$= \frac{1}{2}|\nabla\psi|^2 + |\psi|^2 V(q^k) - \left[\psi^*\left(-\frac{1}{2}\nabla^2\right)\psi + |\psi|^2 V(q^k)\right]$$

$$= \frac{1}{2}\left[|\nabla\psi|^2 + \psi^*\nabla^2\psi\right]$$

which, when

$$\int \psi^*\nabla\psi dV = \int \psi\nabla\psi^* dV$$

as required by the boundary conditions for the case under study, intgrates to

$$\int \frac{1}{4}\nabla^2|\psi|^2 dV \qquad (4.4.3)$$

Thus, the deviations from the mean value, E, of the Hamiltonian density are proportional to the Laplacian of the particle density. Of course the mean square deviation over all space is

$$\frac{1}{4}\left[\int (\nabla^2|\psi|^2)^2 dV\right]^{\frac{1}{2}} \qquad (4.4.4)$$

If 4.4.3 were zero over all space then the solution of the Schrödinger equation would solve the Hamilton-Jacobi equation not just in the mean but everywhere in space: the Hamiltonian-Jacobi equation and the Schrödinger equation would differ only in notation.

For example, in the case of the ground state of the hydrogen atom the mean value of the Hamiltonian density and energy density are both equal to $-1/2$, the deviations from those means are zero for the energy and $\sqrt{10}/2$ for the Hamiltonian density.

The time-independent Schrödinger equation 4.4.2 has a familiar mathematical form (when written in operator notation) from which its formal and manipulational properties are easily abstracted and expressed in a purely algebraic form. Before reviewing these mathematical techniques we must bear in mind that the physical interpretation of Schrödinger's theory which has been given is the physical interpretation of the densities which appear in the Schrödinger Condition; the Schrödinger equation and, a fortiori, the time-independent Schrödinger equation, while dominating the practical applications of the theory, must not be allowed to usurp the central *theoretical* position of the physical interpretation of these probability densities. When we abstract the algebraic structure of the time-independent equation from its (richer) concrete differential form we must have in mind:

- The time-independent equation is only a part of the physical theory

- This abstraction, like any other, produces a structure with considerably less properties than the concrete entity from which it is abstracted.

With these considerations in mind we may write 4.4.1 as

$$\hat{H}\psi = E\psi \qquad (4.4.5)$$

where

$$\hat{H} = -\frac{1}{2}\nabla^2 + V(q^k)$$

for a single-particle system with the obvious generalisation to the many-particle systems.

The boundary conditions which ψ must satisfy (coming from the original Schrödinger condition) require that

$$\int \psi^* \left(-\frac{1}{2}\nabla^2\right)\psi dV = \int \psi \left(-\frac{1}{2}\nabla^2\right)\psi^* dV$$

i.e

$$\int \psi^* \hat{H}\psi dV = \int \psi \hat{H}\psi^* dV$$

Thus 4.4.5 is a special case of the well-documented Hermitian eigenvalue problem and so the formal properties of such systems may be taken over *in toto*.

The salient points are:

- Equation 4.4.5 has an infinite number of solutions (ψ_i, E_i) in which the eigenvalues E_i are all real numbers

- The solutions ψ_i form a complete set for the expansion of any function of the co-ordinates with the same boundary conditions.

- The functions ψ_i form an orthogonal set: for $E_i \neq E_j$

$$\int \psi_i^* \psi_j dV = \int \psi_j^* \psi_i dV = 0$$

- At least some of the ψ_i are square integrable in the sense that

$$\int |\psi_i|^2 dV \ is \ finite$$

The properties of such differential equation are well-known and will not be rehearsed here.

The upshot of this structure being carried by the time-independent Schrödinger equation is that the structure of a vector space may be abstracted from the equation. The ψ_i play the role of vectors in the space and the integrals

$$\int \psi_i^* \psi_j dV$$

the "scalar products" and $\hat{H}$ is a linear operator in this vector space. This vector-space structure is absolutely invaluable in the computational aspects of the Schrödinger theory for conservative systems, allowing the full might of abstract vector-space theory to be brought to bear on practical applications: linear variational methods, perturbation theories etc etc.

But this vector-space structure has no *theoretical* consequences for the Schrödinger theory since the salient points of the meaning and interpretation of the theory lie outside this abstracted structure.

Now we have seen earlier that there are other parts of the theory which have a similar structure to the time-independent equation, principally the expression determining an ensemble of constant momentum per particle:

$$\frac{-i\nabla\psi}{\psi} = \vec{k} \tag{4.4.6}$$

$$or \ \ -i\nabla\psi = \vec{k}\psi \tag{4.4.7}$$

an eigenvalue equation of the same formal structure albeit of a rather simpler type. The question naturally arises: "are there other eigenvalue equations of this type and, if there are, what relationship do they have to the time-independent Schrödinger equation?' The similarity between 4.4.5 and 4.4.6 is striking: both are eigenvalue equations whose eigenvalues are the allowed values of a physical quantity and both have solutions with an underlying vector-space structure. Are there eigenvalue equations which determine the allowed values of other dynamical quantities?

Unfortunately, except in a specialised sense which will be explored in the next chapter, the physical differences between 4.4.5 and 4.4.6 are much greater than their superficial, formal, similarities.

The time-independent Schrödinger equation arises in certain special circumstances and is an equation for one factor of the solution of the dynamical equation in the Schrödinger theory. This equation is generated by a global condition: the equality of the Hamiltonian density and the energy density in the average. Now, in the case of equation 4.4.6 there is no such "history": there is no dynamical equation "behind" 4.4.6, its solution is not a factor in a time-dependent theory; there is no other expression for momentum which would generate an *equation* for the momentum density. In fact as we have seen earlier, equation 4.4.6 is merely the *definition* of an ensemble of constant momentum; it contains no dynamics; it is an *identity*. In fact, without stretching an analogy too far, equation 4.4.6 plays a rather similar role to the velocity identity in Hamiltonian's classical canonical equations:

$$\dot{q}^k = \frac{\partial H}{\partial p_k}$$

is the *definition* of velocity in the canonical theory notwithstanding its striking formal similarity to the dynamical law

$$\dot{p}_k = -\frac{\partial H}{\partial q^k}$$

in the classical canonical theory.

It is now clear that, even if we could generate linear operators which modelled dynamical variables (in the previous section we saw that this is not possible except in trivial cases), only in the case that they obey some dynamical law in the theory is there any chance of their solutions having any substantive (as opposed to formal) similarities with the solutions of the time-independent Schrödinger equation. But the Hamilton-Jacobi equation in its Schrödinger form is the *only* dynamical law which is needed in the Schrödinger theory so that the only real possibility of other equations with the same theoretical status as the time-independent Schrödinger equation are equations which contain the same physical law i.e. transformations of it or other merely mathematical manipulations of the same dynamical law.

Thus the Schrödinger equation even in its time-independent form is the central equation of the theory: other formally similar equations (eigenvalue equations) are either identities defining particular ensembles or equations containing the same dynamical law. The most familiar example of the latter type is the Fourier Transform of the Schrödinger equation; the Schrödinger equation in "momentum space".

Now the abstract vector space spanned by the solutions of the time-independent equation will, course, admit any number of operators in addition to $\hat{H}$. These operators, if Hermitian, will generate complete sets of eigenfunctions in the familiar way and so, in the theory, generate models of ensembles. But only those ensembles generated by $\hat{H}$ (or an operator containing $\hat{H}$ in some way) will be ensembles which satisfy the dynamical law. There is no general law for an arbitrary operator and time development; only the Hamilton-Jacobi-Schrödinger law. It is this simple fact which makes equation 4.4.1 physically and theoretically unique if mathematically commonplace.

The classical-mechanical theory has methods of systematically generating the "constants of the motion"from the dynamical law; methods of identifying those dynamical quantities which are constant throughout the development of the system in time. These methods are associated with the name of Emmy Noether and are used to relate conserved dynamical quantities to the symmetries of the mechanical system. There is no new physics here; just an elegant mathematical technique for elucidating and laying bare the physics which is taking place. Now in the Schrödinger theory there must be some equivalent technique: if the Hamiltonian density is constructed from a ψ which satisfies Schrödinger's equation then it must contain the physics of the "constants of the motion"since they are determined by that motion. These densities are of some theoretical and practical interest and will be examined in the next chapter.

4.5   Conclusions

The conclusions of sections 4.3 and 4.4 are novel and run counter to almost all established thinking on the Schrödinger theory. The conclusions themselves are entirely physical and theoretical; they do not touch the manipulational and practical applications of the theory in the computation of mean values. However they all flow from the central idea that the essence of the Schrödinger theory is the validity of the Hamilton-Jacobi equation on the average over time and space.

Chapter 5

Equations and Identities

5.1 Introduction

The relationship between the condition that an ensemble be homogeneous in some dynamical quantity A (the density per particle of A be constant) and the satisfaction of the dynamical law by that ensemble is obviously easiest to investigate in the case of free particles where the dynamical law is at its simplest and the intuitive picture of the constituents of the ensemble is most clear.

The simplest non-trivial case of a homogeneous ensemble is, as we have seen, the case of a single particle with constant canonical momentum density per particle:

$$-\frac{i}{\psi}\frac{\partial \psi}{\partial q^j} = k_j \tag{5.1.1}$$

i.e.

$$\frac{\partial \psi}{\partial q^j} = ik_j \psi$$

The solution of this elementary equation obviously only fixes the dependence of ψ on *one* of the co-ordinates and, if the system is an ensemble of free single particles, then there are three equations like 5.1.1 —one for each spatial co-ordinate — and a similar equation fixing homogeneities in the "momentum" conjugate to time : energy

$$\frac{\partial \psi}{\partial t} = -iE\psi \tag{5.1.2}$$

It is worth noting here that, in a non-relativistic Schrödinger theory, the identities 5.1.1 and 5.1.2 are completely independent of one another. There is, for example, no requirement that

$$k_1^2 + k_2^2 + k_3^2 = E$$

from this source. The homogeneity condition on energy no more fixes the allowed values of homogeneous momentum than one component of spatial momentum fixes another; it is the dynamical law which establishes this equation not the homogeneity identities.

Now, intuitively, a system which has constant momentum per particle over all space should have constant kinetic energy over all space and, if its motion is subject to no potential energy constraints, its total energy is its kinetic energy and so such an ensemble has constant energy per particle over all space. Thus the Hamiltonian density is equal to the energy density over all space and the Schrödinger condition, requiring only the equality of their *mean values* is trivially satisfied.

Thus, free particles with constant momentum density per particle satisfy the dynamical law; an entirely unsurprising result.

The converse, however, is not true. An ensemble of single-particle systems which has constant kinetic energy density per particle is associated with a solution of the equation

$$\frac{1}{2}\frac{|\nabla\psi|^2}{|\psi|^2} = c \quad \text{(say)} \tag{5.1.3}$$

i.e.

$$\frac{1}{2}|\nabla\psi|^2 = c|\psi|^2$$

which has more solutions than 5.1.1.

Again, intuitively this is obvious since an ensemble of fixed kinetic energy per particle may be composed of members with momentum components in *any direction* and in any combination provided that the sum of the squares (in Cartesians for simplicity!) of the momentum components are all equal.

Now for ensembles of free single-particle systems the condition of constant kinetic energy density per particle (like the condition of constant momentum density per particle) is much stronger than Schrödinger's dynamical law requiring the mean value of the kinetic energy to be a constant. The dynamical law for free single-particle ensemble is, of course,

$$-\frac{1}{2}\nabla^2\psi = E\psi \tag{5.1.4}$$

plus the oscillatory exponential time factor and the familiar boundary conditions from the variational problem. The solutions of 5.1.4 should not necessarily therefore solve 5.1.3. This indeed proves to be the case. The real exponential solutions of 5.1.3 (which, in fact, are of little physical interest) do solve 5.1.2 but the more physically interesting trigonometric solutions of 5.1.4 do not solve 5.1.3 since in these solutions the kinetic energy density per particle is not constant but oscillatory.

It is clear, therefore, that at least in one case, ensembles of constant momentum density do satisfy the dynamical law of Schrödinger's theory. The natural question which now arises is "can momenta be found for which, even in the presence of sources of potential energy, ensembles satisfying the dynamical law will be homogeneous?" Now finding *canonical* momenta which satisfy this condition is tantamount to finding the *co-ordinates* to which the canonical momenta are conjugate. In classical mechanics this can always be done; indeed this is the route by which the Hamilton-Jacobi equation is obtained. But in Schrödinger's mechanics we are much more limited; there is nothing in Schrödinger's theory which corresponds to the freedom, in classical mechanics, to define new co-ordinates in terms of the old co-ordinates *and* momenta. In effect this freedom has already been "used up" in forming the Schrödinger Condition from the Hamilton-Jacobi equation.

Thus, if such momenta are to be found we are limited to transformations of co-ordinates in the strict sense to generate possible new canonical momenta which may be of constant density per particle in ensembles satisfying the dynamical law. However, if we bear in mind the way in which "constants of the motion" are identified in classical mechanics it is easy to reach the provisional opinion that a systematic search for such momenta is doomed to failure or, at least, is not unambiguous.

Apart from accidently-discovered (i.e. intuitively obvious) constant momenta, the constants of the motion in classical mechanics are generated from the variational formulation: Hamilton's principle. The vanishing of the Poisson Bracket provides a diagnostic test for constancy but not a *generator* of such constants. Now in Schrödinger's theory it is likely that a similar condition obtains: the variation principle we have called the Schrödinger Condition is the basic dynamical law and so should generate the details of the motion including those dynamical variables which play the role that "constants of the motion" do in classical mechanics. But the Schrödinger Condition relates only the mean values of the Hamiltonian density and the energy; it is therefore extremely likely that any results obtained from the Schrödinger Condition will only convey information about the *mean values* of candidates for the role of "constants of the motion". Indeed the prime example of such a constant — the Hamiltonian of a conservative system — proves to be just such a case. The Hamiltonian is a classical constant of the motion but in Schrödinger's theory the Hamiltonian density (and the Hamiltonian density per particle) is not a constant but its mean value is equal to the (constant) energy.

However, before continuing with these considerations it is as well to remind ourselves of the consequences of the fact of the two theories — classical mechanics and Schrödinger's mechanics — having different *referents*. The classical mechanics of a single particle is about the dynamics of a single particle but the "quantum mechanics of a single particle"is about ensembles of single-particle systems subject to the Schrödinger Condition.

Now, in the case of a conservative system the particle distributions are independent of time and so the mean value of any dynamical variable which itself contains no explicit time dependence must be independent of time i.e. a constant. For such systems

$$< A >= \int |\psi|^2 A(q^k, -\frac{i}{\psi}\frac{\partial\psi}{\partial q^k})dV = \text{ constant}$$

For example any co-ordinate $A = q^k$ has a mean value $\bar{q}^k$ (say) independent of time. So, *mean values* of any momenta etc. are all constants whereas in the classical mechanical system the individual momenta may well not be constants. Thus to have anything approaching the same physical meaning as in the classical case we must fall back on the original approach; ensembles which are homogeneous for some variable; the density per particle is constant. The question is "how to generate the representative $\psi's$ for such ensembles in which the dynamical law is satisfied?"

We must return to the variational Schrödinger Condition and its consequence the Schrödinger equation. If the Schrödinger equation generates representatives of ensembles which satisfy the dynamical law in (e.g) three dimensional space then any *factorisation* of those solutions should represent ways in which ensembles may be chosen in one or two dimensions which satisfy the dynamical law and which may be homogeneous in some dynamical variable of fewer than three dimensions.

5.2   Separation of the Schrödinger Equation

In this section attention is confined to the dynamics of a particle subject to a conservative field of force in three dimensions. For such systems the Schrödinger equation separates into a simple time dependent equation

$$i\frac{\partial\Psi}{\partial t} = E\Psi \tag{5.2.1}$$

and the time-independent equation

$$\hat{H}\psi = E\psi \tag{5.2.2}$$

substitution of

$$\Psi = \psi \exp\left(-iEt\right)$$

gives the space component as the essential solution of 5.2.2

$$\hat{H}\psi = E\psi \qquad (5.2.3)$$

Now, considered purely as a partial differential equation, 5.2.3 may be solved by any of the known techniques. In particular the most powerful analytical tool for the solution of partial differential equations - solution by separation of variables - may be used.

The techniques of finding co-ordinate systems which separate 5.2.3 and enable it to be solved by the techniques available for ordinary differential equations are well-known and well-developed. The central idea is, approximately, that if a co-ordinate system can be found in which $\hat{H}$ breaks up into a *sum* of operators depending on one or two co-ordinates each, then the solution of 5.2.3 is the product of the solution of the eigenvalue equation associated with these operators. For a complete solution of the three-dimensional equation, $\hat{H}$ must be broken up, by suitable choice of co-ordinate system, into a sum of three operators, each depending on only one of the chosen co-ordinates.

From the point of view of the project in hand - the finding of homogeneous ensembles - the separation of the Schrödinger *equation* does not look too optimistic for two related reasons:

- The operator $\hat{H}$ is a quadratic form in the momentum components - i.e the separation is likely to be into operators related to kinetic energy components rather than momenta.

- The operator $\hat{H}$ does not generate the Hamiltonian density - it only generates its mean value when integrated - therefore the physical interpretation of the results of the separation technique must be done with caution.

With these reservations in mind we press on with the technique.

The central idea is, that if the solution of 5.2.3 can be expressed as a product

$$\psi(q^1, q^2, q^3) = u(q^1)v(q^2)w(q^3) \qquad (5.2.4)$$

for some co-ordinate system (q^1, q^2, q^3) then the one-dimensional functions satisfy the equations

$$\hat{U}u(q^1) = \ell u(q^1) \qquad (5.2.5)$$

$$\hat{V}v(q^2) = mv(q^2) \qquad (5.2.6)$$

$$\hat{W}w(q^3) = nw(q^3) \qquad (5.2.7)$$

for constants ℓ, m, and n and further $\hat{U}$, $\hat{V}$, $\hat{W}$ have the property of *commuting* with $\hat{H}$ in the sense that, for any function $f(q^1, q^2, q^3)$ satisfying the boundary conditions of the problem

$$\hat{H}\hat{U}f(q^1) = \hat{U}\hat{H}f(q^1) \quad etc.$$

The essence of the proof is that $\hat{H}$ is (apart from some algebraic factors) the *sum* of the operators $\hat{U}$, $\hat{V}$ and $\hat{W}$ and since

$$(\hat{H})\hat{H}f(q^1) = \hat{H}(\hat{H}f(q^1))$$

either $\hat{H}$ may be expressed in terms of $\hat{U}$, $\hat{V}$ and $\hat{W}$.

In practice since each of $\hat{U}$, $\hat{V}$ and $\hat{W}$ involve *only one* of the q^k, the commutation in the other two is trivial and this fact may be used to place 5.2.3 and 5.2.5 together in a *formally* equivalent setting

$$\hat{H}\psi = E\psi \tag{5.2.8}$$
$$\hat{U}\psi = \ell\psi \tag{5.2.9}$$
$$\hat{V}\psi = m\psi \tag{5.2.10}$$
$$\hat{W}\psi = n\psi \tag{5.2.11}$$

But, of course, it is the first of these equations which *generates* the other three, the similarity is only a formal one; in no sense can the equations involving $\hat{U}$, $\hat{V}$ and $\hat{W}$ be considered of an equal *theoretical* status to those involving $\hat{H}$.

However these equations are reminiscent of the equation defining a homogeneous ensemble i.e if $\psi^*\hat{U}\psi$ is the U-density then the U-density per particle is

$$\frac{1}{|\psi|^2}\psi^*\hat{U}\psi$$

which is constant if

$$\hat{U}\psi = \ell\psi$$

Thus *if* the operators $\hat{U}$, $\hat{V}$, $\hat{W}$ correspond to physical quantities in the sense that

$$\psi^*\hat{U}\psi = |\psi|^2 U(q^k, -\frac{i}{\psi}\frac{\partial\psi}{\partial q^k})$$

for a classical variable U then the ensemble represented by ψ is homogeneous in that variable. Now as we noted above, the operators $\hat{U}$, $\hat{V}$, $\hat{W}$ should be additive components of $\hat{H}$ and as such ought, in general, to be composed of a component of the Laplacian plus possibly, a term from the de-composition of the potential energy expression in the co-ordinate system (q^1, q^2, q^3).

However, they are generated from $\hat{H}$ which has the Laplacian as "kinetic energy operator" and so, in general, we might expect that these linear operators only represent components of kinetic energy in the same sense that the Laplacian represents total kinetic energy: in the mean and not locally. In general, $\psi^* \hat{U} \psi$ will not be the density of a kinetic energy component in the co-ordinate q^1 since this is known to be q^1 component of $|\nabla \psi|^2/2$

It may happen that there is more than one co-ordinate system in which this separation is possible which may or may not have separation operators in common. A familiar example is the Schrödinger equation for the (fixed nucleus) hydrogen atom which separates in no less than four co-ordinate systems generating a total of twelve separation operators, not all of which are different. The relationship between the set of all distinct separation operators and the symmetry of the Schrödinger equation is well-known in the theory of partial differential equations and will not be pursued here since we are primarily concerned with the physical interpretation of the separation process.

It is not possible to give a *general* statement of the relationship between the separation operators in a particular co-ordinate system and the canonical momenta in those co-ordinates since the metric coefficients take on different forms in each particular case. If we confine attention to the familiar 11 orthogonal co-ordinate systems of second degree it is easy to state the conditions under which the Schrödinger equation will separate; since the Laplacian separates in all of these systems the conditions are on the *form* of the potential energy function. In all but three of these systems (sphero-conal, ellipsoidal and parabaloidal) at least one of the separation operators has a contribution from the Laplacian of the simple form

$$\frac{\partial^2}{\partial q^{k^2}}$$

meaning that, if the potential energy function is independent of q^k, the eigenvalue equation has the form

$$\frac{d^2 f}{dq^{k^2}} + \lambda f = 0$$

with the familiar trigonometric solutions. In these cases one may take the complex conjugate pairs of exponential solutions of

$$\pm i \frac{df(q^k)}{dq^k} + \nu f(q^k) = 0$$

as solutions of the original equation and so generate momentum-homogeneous solutions of the original Schrödinger equation.

In the more general case the separated Laplacian component takes is full form

$$\frac{1}{h_k}\frac{d}{dq^k}\frac{h_1 h_2 h_3}{h_k}\frac{df}{dq^k}$$

and no decomposition into a momentum eigenfunction equation is possible.

We may therefore conjecture that a given Hamiltonian density d_H which, through the Schrödinger Condition, generates a particular Hamiltonian operator $\hat{H}$ and also determines all the "constants of the motion" for the system. That is this density and the dynamical law fixes all those homogeneous ensembles which are *compatible* with the dynamics; whose homogeneity is not disturbed by the unfolding of the system's dynamics. Apart from physically trivial extensions like multiples and powers of $\hat{H}$ and the like, the operators determining these homogeneous ensembles are just the *separation operators* of the Schrödinger equation in various co-ordinate systems. Again, such sets of separation operators, their scalar multiples, products and powers will generate a set of additional operators (a group, usually!) which is often of mathematical interest but which contains no new dynamical information; Schrödinger's dynamical *law* determines all these quantities.

5.3   An example of Separation

The Hamiltonian operator for the (fixed nucleus) hydrogen atom in the electrostatic approximation is

$$-\frac{1}{2}\nabla^2 - \frac{1}{r} = \hat{H} \tag{5.3.1}$$

where r is the electron-nucleus distance. This operator and the associated time-independent Schrödinger equation separates in four co-ordinate systems; of which the most familiar is the spherical polar system (r, θ, ϕ). Writing the full solution as $\psi(r, \theta, \phi)$, the separation is accomplished by

$$\psi(r, \theta, \phi) = R(r)\Theta(\theta)\Phi(\phi)$$

with the separate equations forming a hierarchy:

$$\frac{d^2\Phi}{d\phi^2} = \alpha\Phi \tag{5.3.2}$$

$$\frac{1}{\sin\theta}\frac{d}{d\theta}\sin\theta\frac{d\Theta}{d\theta} - \frac{\alpha}{\sin^2\theta} = \beta\Theta$$

$$-\frac{1}{r^2}\frac{d}{dr}\left(r^2\frac{dR}{dr}\right) + \left(-\frac{1}{r} - \frac{\beta}{r^2}\right)R = ER$$

The interpretation is fairly straightforward particularly since the last of the equations is, effectively, the full Schrödinger equation in a parametrical form. The first term of the last equation is the representative of the "radial kinetic energy", the term β/r^2 (recall the unit of mass is absent by convention) is β/I the angular kinetic energy, (I is the moment of inertia of a particle of distance r from the origin). In the second equation the left hand side is the representative of angular kinetic energy (apart from the angle-independent moment of inertia) while α is the "ϕ- component"of the angular kinetic energy available from the first equation; $r^2\sin^2\theta$ is the moment of inertia of particle of unit mass distance r from the origin projected onto the $x - y$ plane. This latter equation gives preference to the $x - y$ plane simply from the definition of spherical polar co-ordinates.

The interesting equation is the first one since, from our point of view, it clearly has some solutions in common with

$$\frac{d\Phi}{d\phi} = \lambda\Phi$$

which defines a ensemble homogeneous in momentum conjugate to ϕ. However, this is in some sense an "accident"as we can see from the θ equation which, even in the absence of the α term (i.e. when $\alpha = 0$) does not define an ensemble homogeneous in momentum conjugate to θ. Both θ and ϕ terms are components of the Laplacian which gives the correct mean value of kinetic energy not the correct kinetic energy density and, *a fortiori*, not the correct momentum densities.

In sphero-conal co-ordinates the angular kinetic energy operator also effects a separation so that the last of the three equations is repeated in this co-ordinate system. However, in sphero-conals there is dependence of the range of the co-ordinates on two parameters b and c which fix the eccentricity of the elliptic conical co-ordinate surfaces. This parametric dependence of the co-ordinate surfaces generates whole families of separation operators which differ from one another only in the relative weighting given to components of the angular kinetic energy and, in general, no momentum-homogeneous ensemble is generated.

5.4   Relationship to Other Dynamical Variables

In an earlier chapter we showed that the representation of the density of a dynamical variable A by an expression

$$\psi^* \hat{A} \psi \qquad (5.4.1)$$

for a linear operator $\hat{A}$ is, in general, impossible. Only when $\hat{A}$ is *linear* in momenta is such a representation guaranteed; i.e. in effect, only the momentum densities themselves are expressible in this form. Indeed in the most important case of the kinetic energy it is quite straight forward to show that no momentum operators will satisfy both the Poisson Bracket relationships and generate a linear operator for the kinetic energy density.

But, in the special case of the representatives ψ of ensembles which satisfy the dynamical law, the mean value of the kinetic energy can be generated by the integration of a density of the form 5.4.1. This condition obtains because the variational solution of the Schrödinger condition replaces the kinetic energy density by a densities of the form 5.4.1 plus some boundary conditions ensuring the equality of the mean value of the kinetic energy and the replacement operator - the Laplacian.

The decomposition of the Schrödinger equation by separation discussed in the last section provides components of the Laplacian which have the same properties as the Laplacian in their relationship to the components of kinetic energy. This separation technique provides a set of *linear operators* which represent dynamical variables in the same "mean value" sense that the Hamiltonian operator represents the Hamiltonian density.

If we ask for a linear operator which represents an *arbitrary* dynamical variable in the "mean value" sense we are at a loss to know how to generate such an operator. Indeed such operators cannot exist in general since even for those operators generated from the dynamical law, it was the *boundary conditions* which ensured that the mean values were the correct ones. Thus, the formation of linear operators to generate even the correct mean values of arbitrary dynamical variables is no more possible than the generation of the correct densities by expressions of the form 5.4.1.

The linear operators in the Schrödinger theory fall into three classes:

1. The "momentum operators" $-i\partial/\partial q^k$; a solution of the eigenvalue equation

$$-i\frac{\partial \phi}{\partial q^k} = \lambda \phi$$

is a function homogeneous in the momentum component; the momentum density per particle is constant over all space.

2. The "Hamiltonian operators"; a solution of the time-independent Schrödinger equation

$$\hat{H}\psi = E\psi$$

has constant energy E which is equal to the mean value of the Hamiltonian density. A solution of the full Schrödinger equation

$$\hat{H}\Psi = i\frac{\partial\Psi}{\partial t}$$

also has equality between energy and Hamiltonian densities but both are mean values.

3. The operators obtained by separation of the Schrödinger equation (if this is possible): $\hat{U}$ then a solution of

$$\hat{U}f = \ell f$$

is a factor of the solution of the Schrödinger equation. In some cases (3) may be equivalent to (1), in all cases the eigenvalue of $\hat{U}$ is a "constant of the motion"— usually an additive component of the total energy E.

Clearly (3) is completely derived from (2) and the set of $\hat{U}$s may be empty for some potentials.

The distinction between (1) and (2) is that (2) is an equation - it embodies the dynamical law of Schrödinger's mechanics - whereas (1) is just a set of identities - defining certain homogeneous ensembles independent of any dynamical law.

It is therefore worth conjecturing that the only linear operators in the Schrödinger theory which "represent"dynamical variables are the momentum operators derived by abstraction from the Hamilton-Jacobi-Schrödinger momentum densities and the Hamiltonian operator (plus possibly the components of the Hamiltonian operator for particular systems). The momentum operators represent the momenta in the sense that

$$|\psi|^2\left(-\frac{i}{\psi}\frac{\partial\psi}{\partial q^k}\right) = \psi^*\left(-i\frac{\partial\psi}{\partial q^k}\right) = \psi^*\hat{P}_k\psi$$

is the momentum density while the Hamiltonian operator (and its components, if any) represent the Hamiltonian in the sense that the mean value of the Hamiltonian density is

$$\int\psi^*\hat{H}\psi dV$$

The actual Hamiltonian density is, of course,

$$|\psi|^2 H(q^k, -\frac{i}{\psi}\frac{\partial \psi}{\partial q^k}) = \frac{1}{2}|\nabla \psi|^2 + V(q^k)|\psi|^2$$

Chapter 6

A Posteriori **Connections**

6.1 Introduction

We have used the Schrödinger Condition to make the transition from Classical Particle Mechanics (the Hamilton-Jacobi equation) to Quantum Mechanics (the Schrödinger equation) where, in both cases, the co-ordinates q^i and t are a set of $3N + 1$ *independent variables*. In both cases the *trajectories* of the particles are quite remote from the formulation of the dynamical law. In the case of the Hamilton-Jacobi equation the trajectories are characteristics (normals) to S the solution of the equation and in the case of Schrödinger's equation, the individual trajectories are not determined at all; only mean values are fixed.

Having made this transition and having made clear the *physical* difference between classical and quantum mechanics, the question which naturally arises is:

> "Can the *quantum mechanical* dynamical law be cast into a form reminiscent of the earlier and more basic forms of classical mechanics — the Lagrange equations or even the equations of Newton?"

That is, can we *reverse* the chain of reasoning given above and generate some analogue of the familiar expression of the particle trajectories of classical mechanics in which the co-ordinates q^i appear as explicit functions of t; a form in which the imposition of the dynamical law generates $q^i(t)$ rather than independent q^i and t? In quantum mechanics what replaces the curve in $3N$-space *parametrised* by t?

Any answer which we are able to provide to these questions must be tempered by the knowledge that only *mean values* of q^i and p_i etc. are

available from Schrödinger's mechanics. With this proviso, we can take
the appropriate steps and generate other formulations of quantum mechanics which stress the similarities between classical and quantum equations
other than the Hamilton-Jacobi/Schrödinger connection; formulations due
to Dirac and Heisenberg. It must also be said at the outset that, if we seek
a formulation of quantum mechanics which is parametrised by t, then it is
only sensible to do so for non-stationary states; systems whose probability
densities do, in fact, *change* with time.

In this chapter, in contrast to the rest of the work, we include the unit
of action explicitly ($k = \hbar$) and also explicitly write the mass as m to act
as a reminder of the relationship with classical mechanics.

6.2   Evolution of Mean Values

The probability distribution of a canonical co-ordinate q^i and a canonical
momentum component p_i are given (respectively) by

$$|\Psi|^2 q^i = \Psi^* q^i \Psi$$

and

$$-|\Psi|^2 \frac{\hbar i}{\Psi} \frac{\partial \Psi}{\partial q^i} = \Psi^* \hat{p}_i \Psi \quad \text{(say)}$$

Both of these expressions have the general form

$$\Psi^* \hat{A} \Psi$$

The mean value of quantities of this kind (for normalised Ψ) is

$$< \hat{A} >_\Psi = \int \Psi^* \hat{A} \Psi dV = < \hat{A} > \quad \text{(say)}$$

if we restict attention to just one Ψ.

Now, $< \hat{A} >$ is a function of t only (since integration has been carried
out over the q^i) so that:

$$\frac{d < \hat{A} >}{dt} = \frac{\partial < \hat{A} >}{\partial t}$$

$$= \int \Psi^* \left(\frac{\partial \hat{A}}{\partial t}\right) \Psi dV + \int \left(\frac{\partial \Psi^*}{\partial t}\right) \hat{A}\Psi dV + \int \Psi^* \hat{A} \left(\frac{\partial \Psi}{\partial t}\right) dV$$

If now Ψ is, in fact, a solution of the Schrödinger equation:

$$\hat{H}\Psi = i\hbar \frac{\partial \Psi}{\partial t} \tag{6.2.1}$$

$$\hat{H}\Psi^* = -i\hbar\frac{\partial\Psi^*}{\partial t}$$

since Ψ is constrained by the boundary conditions to make $\hat{H}$ Hermitian.

Thus, using these substitutions for Ψ and Ψ^*, we obtain an expression for the time evolution of the mean value of $\hat{A}$ over a probability distribution whose associated Ψ solves the Schrödinger equation; i.e. over a probability distribution satisfying the dynamical law of Schrödinger's mechanics:

$$\frac{d<\hat{A}>}{dt} = \frac{\partial <\hat{A}>}{\partial t} = \left\langle\frac{\partial\hat{A}}{\partial t}\right\rangle + \frac{i}{\hbar}\left\langle\hat{H}\hat{A} - \hat{A}\hat{H}\right\rangle \tag{6.2.2}$$

Since the function Ψ has been assumed to solve the Schrödinger equation, this equation is no longer an identity; it is an equation and is only true for those motions which are governed by quantum mechanics.

In fact, 6.2.2 can be generalised slightly to hold for the time development of

$$A_{ij} = \int \Psi_i^* \hat{A}\Psi^j\, dV$$

where both Ψ_i and Ψ_j solve a Schrödinger equation — repeating the whole procedure gives

$$\frac{dA_{ij}}{dt} = \frac{\partial A_{ij}}{\partial t} = \left(\frac{\partial A}{\partial t}\right)_{ij} + \frac{i}{\hbar}\left(\hat{H}\hat{A} - \hat{A}\hat{H}\right)_{ij} \tag{6.2.3}$$

or, in matrix notation,

$$\frac{dA}{dt} = \acute{A} + \frac{i}{\hbar}\left(HA - AH\right) \tag{6.2.4}$$

where $\acute{A}$ is the matrix of

$$\acute{A}_{ij} = \int \Psi_i^* \left(\frac{\partial\hat{A}}{\partial t}\right)\Psi_j\, dV = \acute{A}_{ji}$$

The notation of the right-hand-side is rather ambiguous; it means the matrix of the commutator not the commutator of the matrices. This equation is often written as

$$\frac{dA}{dt} = \frac{\partial A}{\partial t} + \frac{i}{\hbar}\left(HA - AH\right) \tag{6.2.5}$$

to emphasise certain similarities with the classical mechanical equations. However, it must be noted that the two derivative terms are written with scant regard for precision in this form.

6.3   Dirac and Heisenberg's Formulation

In classical mechanics, a dynamical quantity A will, generally, depend on the co-ordinates q^i, the momenta p_i and t:

$$A = A(q^i, p_i, t)$$

so that the total time derivative of such a quantity is

$$\frac{dA}{dt} = \frac{\partial A}{\partial t} + \sum_{i=1}^{3N} \left(\frac{\partial a}{\partial q^i} \dot{q}^i + \frac{\partial A}{\partial p_i} \dot{p}_i \right)$$

If now the $\dot{q}^i$ and the $\dot{p}_i$ are such that Hamilton's equations of motion are satisfied, then for values of A which occur during *real motion* of the dynamical system we have

$$\frac{dA}{dt} = \frac{\partial A}{\partial t} + \sum_{i=1}^{3N} \left(\frac{\partial A}{\partial q^i} \frac{\partial H}{\partial p_i} - \frac{\partial H}{\partial q^i} \frac{\partial A}{\partial p_i} \right) \tag{6.3.1}$$

The summation on the right of 6.3.1 has been met before — it is the Poisson Bracket of A and H: $[A, H]_{p,q}$ which is independent of the particular co-ordinates q^i and momenta p_i and may be written simply as $[A, H]$. Equation 6.3.1 may therefore be written

$$\frac{dA}{dt} = \frac{\partial A}{\partial t} + [A, H] \tag{6.3.2}$$

and the similarity of structure between this result and 6.2.5 is striking; the Poisson Bracket in 6.3.2 is replaced by the commutator of two matrices to form 6.2.5. Dirac has made this a prescription for the transition from classical mechanics to the corresponding quantum mechanical equation. That is, 6.2.5 is not taken as a consequence of an identity and some mean values over functions which satisfy Schrödinger's equation, but is taken as the *starting point* of a formulation of quantum theory.

The question now arises:

> "How does one actually calculate dynamical quantities from 6.2.5 and can one infer from 6.2.5 the existence of the functions Ψ_k which *we* used to generate the equation?"

As it stands, 6.2.5 is a matrix equation. If, as is very often the case, the quantity A has no explicit time dependence then 6.2.5 is an *ordinary* differential equation for the time development of the elements of a matrix.

It is clear that the time development of a mean value will not be sufficient to generate the functions Ψ_k.

However, it is also clear that the solutions of 6.2.5 will be of some value in quantum theory. Equation 6.2.5 is now universally known as Heisenberg's Equation. The most common applications are just those for which A has no explicit time dependence and so, in these cases,

$$\frac{dA}{dt} = \frac{i}{\hbar}[A, H]$$

where the square brackets are now understood to mean

$$[A, H] = (AH - HA)$$

for the *matrix* of the commutator of A and H.

In deriving 6.2.5 we specifically assumed that the density of A was (quantum-mechanically) able to be expressed as $\Psi^* \hat{A} \Psi$ simply on the basis of this form for the q^i and p_i. We have already seen that, unless we are prepared to admit the existence of a privileged co-ordinate frame (Cartesians), there are many classical A's which do not admit this quantum description. Thus, we will investigate the consequences of Heisenberg's equation for these simplest cases in the first instance.

6.4 Heisenberg's Equation for Q and P

Taking the simplest possible application of 6.2.4, we set $A = q$ and use the symbol Q for the matrix

$$Q_{ij} = \int \Psi_i^* q \Psi_j \, dV$$

so that equation 6.2.4 becomes

$$\dot{Q} = \frac{dQ}{dt} = \frac{i}{\hbar}[HQ - QH]$$

since $\acute{Q} = 0$ by hypothesis; the co-ordinates q^i, the momenta p_i and time t are the *independent* variables used to describe the dynamics — $\partial q / \partial t \equiv 0$. If the Hamiltonian operator is a simple case:

$$\hat{H} = -\frac{\hbar^2}{2m}\nabla^2 + V(q)$$

then, working in Cartesians but retaining the general notation Q, we have

$$[HQ - QH] = \frac{1}{m}P$$

so that

$$\dot{Q} = \frac{dQ}{dt} = \frac{1}{m}P = \frac{P}{m}$$

the familiar relationship between "velocity" and "momentum", which has an entirely classical form.

This result is independent of the form of V, indeed it is independent of the very existence of V and so it cannot be a *dynamical* equation; it is, of course, analogous to the corresponding identity in classical mechanics.

However, the choice of $A = p$ yields a genuine dynamical equation depending on the form of V, the law of force under which the particle moves:

$$\dot{P} = \frac{dP}{dt} = \frac{i}{\hbar}[HP - PH]$$

Here, the matrix of the commutator *does* depend on the law of force operating since the differential form of any momentum operator $\hat{p}$ corresponding to a classical p, even if it commutes with the ∇^2 term in H, has an effect on $V(q)$. For the moment we will defer consideration of a non-zero V and carry forward the analysis for the (dynamically) trivial case of $V \equiv 0$: a free particle. In this case

$$\dot{Q} = \frac{dQ}{dt} = 0$$

(again, in Cartesians to avoid any possible complications with the definition of $\hat{p}$ and ensuring that $\hat{p}$ and ∇^2 commute). Thus we have two relationships which must hold for the motion of a free particle, formally rather similar to the canonical forms of Hamilton:

$$\dot{Q} = \frac{dQ}{dt} = \frac{1}{m}P$$

$$\dot{P} = \frac{dP}{dt} = 0$$

in which the constant $\hbar$ does not appear. As matrix equations these obviously have the solutions

$$P = P^0$$
$$Q = Q^0 + \frac{1}{m}P^0 t$$

where Q^0 and P^0 are constants of integration. The matrices Q^0 and P^0 are the "initial values" of the matrices Q and P at $t = t_0 = 0$ (say). These matrices clearly play the rôle of the initial values of the q^i and p_i in classical mechanics; containing the time t, not as an explicit time dependence of q^i, p_i on t (since these are the $6N + 1$ independent variables) but via the solution of the Schrödinger equation. This fact which is masked by our abbreviated notation for the matrix elements:

$$Q_{ij}^0 = \int \Psi_i(t_0)^* q \Psi_j(t_0) dV$$

$$P_{ij}^0 = \int \Psi_i(t_0)^* \hat{p} \Psi_j(t_0) dV$$

Thus, in order to proceed any further with the explicit solution of the dynamical problem we would seem to need the solution of the underlying Schrödinger equation at one particular time; having done this the time dependence of the matrices of mean values is fixed by Heisenberg's equation.

In point of fact, by using the quantum commutator analogue of the fundamental Poisson brackets involving the canonical co-ordinates and momenta in the expression for a (time-independent) Hamiltonian, one can generate the energy eigenvalues in full agreement with those of the Schrödinger equation but not, of course, the solutions Ψ_i.

It is worthwhile to note here the rather different form which the dynamical law takes in the Heisenberg ("Newtonian") mean-value relationship from that in the Schrödinger ("Hamilton-Jacobi") differential form. In the Schrödinger form, if one needs to know the mean value of a particular dynamical quantity at time t one solves the time-dependent Schrödinger equation, obtains the function $\Psi(q, t)$ and the evaluation of the mean value

$$\int |\Psi|^2 A(q^i, p_i, t) dV$$

is immediate. The mean value and the associated *distribution* are available from Ψ as functions of t.

In the Heisenberg formulation the time dependence is thrown into the *form* of the equation relating mean values at different times; one cannot get the *distribution* since the integration over this distribution has, effectively, already *been done* at time $t = t_0$. In this case one only generate the way in which the mean value changes with time, not the way in which the associated distributions change with time. Presumably, where mean values at different times are computed by the two approaches and compared, they will agree:

$$Q(t) = \int \Psi(t)^* q \Psi(t) dV$$

where the left-hand-side is computed from the Heisenberg equation and the right-hand-side from the solution of the Schrödinger equation.

In the introduction to this chapter we alleged that it is only meaningful to compute the time dependence of the mean values of quantities which

did, in fact, change with time and yet the example which we have used —
the free particle — has a Schrödinger equation which *separates* into a simple exponential time-dependent factor and a *time-independent* Schrödinger
equation. That is, the particle distributions associated with the free particle Schrödinger equation are *independent of time*! Since we generated a
solution of the Heisenberg equation

$$Q = Q^0 + \frac{P}{m}t \qquad (6.4.1)$$

which is clearly time-dependent, this needs some re-examination.

Solutions of the Schrödinger equation which factorise in this familiar
way all have particle distributions which are independent of time:

$$\Psi^*\Psi = \psi^*\psi$$

This is not inconsistent with the result 6.4.1 since it is easy to show that,
under these circumstances, the mean values of the canonical momentum
components are all *zero* leading to

$$P = P^0 = 0$$

$$Q = Q^0 = \quad \text{constant}$$

Neither P nor Q change with time and our whole theory is apparently
vacuous!

But it is evident on both physical and mathematical grounds what the
source of the difficulty is. Clearly an (unacknowledged) axiom of quantum
mechanics is that particles are always in motion[1]; any individual concrete
free particle moves and therefore has momentum. Now, the solution of
the Schrödinger equation refers to an *ensemble* of all possible motions of
the system with a given energy: an equi-energetic ensemble, say. An equi-
energetic ensemble of free particles has, in fact, zero mean momentum since,
if *all possible* members are present in such an ensemble, for each member of
momentum $p^i = k$ there must be one of momentum $p^i = -k$, equal numbers
in all directions; there being no source of potential to hinder the motion.
If however, we can arrange to create a description of an equi-energetic ensemble *together with* the condition that (say) only left-to-right motion is
allowed, then we would have generated an ensemble to which the solution
of the Heisenberg equation given above might actually seem most applica-
ble; an equi-energetic ensemble which does have non-zero momentum mean
value. One which equation 6.4.1 would predict non-trivial time evolution
of the mean value of the conjugate co-ordinate.

---

[1] "Motion is the mode of existence of matter!"

Mathematically, this amounts to choosing a momentum-homogeneous ensemble of free particles which also solves the Schrödinger equation. These solutions can actually be generated but at the expense of *constraining* the possible motions of the particle; i.e. a restricted ensemble, a specific choice of co-ordinate system. If one works in a co-ordinate system where $\hat{H}$ is the square of some other operator — in our case where the kinetic energy is the simple square of an operator — then the eigenfunctions of the square root of the Hamiltonian are also solutions of the Schrödinger equation. The two operators $\hat{H}$ and $\hat{H}^{1/2}$ between them define a sub-ensemble of the ensemble associated with the eigenfunctions of $\hat{H}$; a *sub*-ensemble because the square root of $\hat{H}$ can not be extracted in general, only in particular co-ordinate systems. This type of problem has been discussed in section 4.2.

Clearly in the free-particle case, use of Cartesian co-ordinates *(again!)* enables the square root to be extracted easily and so the eigenfunctions of the free-particle Hamiltonian are the eigenfunctions of the Cartesian momentum operators. This relationship may be written as

$$\nabla\psi = \vec{k}\psi$$

$$\nabla^2\psi = \nabla \cdot (\nabla\psi) = \nabla \cdot (\vec{k}\psi) = \vec{k} \cdot \nabla\psi = |\vec{k}|^2\psi$$

but this formulation gives an over-general co-ordinate-free impression, it ignores the fact that it is only in the one specific co-ordinate system that *grad* and *div* are "the same"; $\nabla^2 = (\nabla)^2$. In physical terms, an ensemble of free particle systems may only be chosen to be momentum-homogeneous if all the momenta concerned are *linear* momenta. The fact that the Laplacian operator factors into two *"different"* terms in other co-ordinate systems means that free-particle ensembles do exist which are not linear-momentum homogeneous; the familiar Bessel function solutions arising from the separation in spherical polars, for example.

However, notwithstanding all these reservations, the linear momentum eigenfunctions

$$\psi_k = exp\left(\frac{i}{\hbar}\vec{k} \cdot \vec{r}\right) \tag{6.4.2}$$

do satisfy the Schrödinger equation and generate a non-zero P^0.

But now we are in a dilemma, elementary considerations indicate that $< x >$ is independent of time for a time-independent Hamiltonian (the left-hand-side of 6.4.1 is zero) but the right-hand-side contains a linear time dependence this time with non-zero P^0! The solution lies in the fact that the momentum eigenfunctions 6.4.2 cannot be normalised except by limiting them to a finite domain. If,however, the functions are limited to a finite domain the boundary conditions which must be imposed on them (vanishing at the end-points of the region) force the functions to be *real*, representing equal mixtures of momenta k and $-k$, taking us back to a zero P^0 again.

Except in the case that the Hamiltonian operator depends explicitly on t, preventing a separation of the solutions of the Schrödinger equation into an exponential time-dependent factor, the time derivative of all mean values is zero.

It has been stressed in this section that, as an equation of motion for a matrix, the Heisenberg equation has no use for the solutions of the Schrödinger equation except as the provider of the initial conditions of the motion but we can use some purely formal devices to bring the two formulations of quantum theory a little closer. The key relationship between the two schemes is that given earlier in this section, the equality the matrix element at time t in the Heisenberg formulation and the mean value involving a solution of the Schrödinger equation at time t:

$$A(t) = \int \Psi(t)^* \hat{A}\Psi(t)dV \qquad (6.4.3)$$

where, to avoid complications, we assume that A can indeed be represented by $\hat{A}$ and that $\hat{A}$ does not depend on t.

Now, in the solution of the Heisenberg equation for $A(t)$ there will occur a constant of integration A^0) which is the value of 6.4.3 when $t = t_0$:

$$A^0 = A(t_0) = \int \Psi(t_0)^* \hat{A}\Psi(t_0)dV$$

We can incorporate the time development *under* the integral sign if we assume the existence of some formal operator which generates $\Psi(t)$ from $\Psi(t_0)$:

$$\Psi(t) = \hat{U}(t_0,t)\Psi(t_0) \quad \hat{U}(t_0,t_0) = 1$$

it is easy to see that such a formal operator, if it exists, must be unitary and also, incidentally, that such a $\hat{U}$ will be formally closely related to $\hat{H}$. In this case

$$A(t) = \int \Psi(t_0)^* \left(\hat{U}^\dagger \hat{A}\hat{U}\right) \Psi(t_0)dV$$

where we may now think of the time development of the mean value A as being the mean value of the operator $(\hat{U}^\dagger \hat{A}\hat{U})$ over the distribution associated with the solution of the Schrödinger equation at $t = t_0$.[2] It

---

[2] This formal device has led to some confusions about the all-important *interpretation* of quantum mechanics. It is sometimes seen as a mystery why "all information about the evolution of a system can be contained in the solution of the Schrödinger equation at a *single* instant of time". But this is true of *all* initial conditions. In classical mechanics, knowledge of $q^i(t_0)$ and $p_i(t_0)$ fixes all future $q^i(t)$ and $p_i(t)$ *provided* the law of force is known and the dynamical law is assumed.

must be stressed that this description, although very common, is, to say the very least, a rather eccentric way of incorporating the constants of integration of an ordinary differential equation into the general solution. It tends to disguise the very straightforward links between the Schrödinger and Heisenberg schemes and does not show the clear relationship between the evolution of ensemble mean values in quantum theory and the evolution of individual systems in classical mechanics.

6.5   The Ehrenfest Relations

In the last section we looked at the Heisenberg equation of motion for the simplest possible system, we can now explore the relationship with classical mechanics a little further and consider a particle moving in a (conservative) field of force i.e. with Hamiltonian

$$\hat{H} = -\frac{\hbar^2}{2m}\nabla^2 + V(q^i)$$

For the moment, working in Cartesian co-ordinates, we can easily see that the equation for the evolution of the mean value of one of the co-ordinates is

$$\frac{d}{dt} < q^i > = \frac{1}{m} < \hat{p}_i >$$ (6.5.1)

as usual and, if we assume the Cartesian form for the momentum operator, $\hat{p}_i = -i\hbar\partial/\partial q^i$

$$\frac{d}{dt} < \hat{p}_i > = -\left\langle \frac{\partial V}{\partial q^i} \right\rangle$$ (6.5.2)

Obviously, we may differentiate 6.5.1 once to yield

$$\frac{d^2}{dt^2} < q^i > = \frac{1}{m}\frac{d}{dt} < \hat{p}_i >$$

and, using 6.5.2, we have

$$m\frac{d^2}{dt^2} < q^i > = -\left\langle \frac{\partial V}{\partial q^i} \right\rangle$$ (6.5.3)

for each of the three Cartesian co-ordinates; $q^i = x, y, z$. In the spirit of vectorial mechanics, using

$$\vec{r} = x\vec{i} + y\vec{j} + z\vec{k}$$

we may summarise these three relationships as

$$m\frac{d^2}{dt^2} < \vec{r} > = -\left\langle \nabla V(x, y, z) \right\rangle$$ (6.5.4)

an equation very reminiscent of Newton's famous law: it is the so-called Ehrenfest relationship.

This whole development has been carried through in Cartesian coordinates for reasons which are, by now, very obvious; the result hinges on the fact that, in Cartesians, the quantum-mechanical momentum operator and the "classical" gradient operator are *the same* (apart from a numerical factor). What is generated by the equations is, in general, a numerical multiple of $\hat{p}_i V$ which may or may not be a component of ∇V depending on ones choice of "momentum operator". These problems are addressed to some extent in Appendix C. As we sketch in that appendix, there is in fact no way of choosing a definition of canonical momentum operators $\hat{p}_i$, conjugate to co-ordinates q^i, which satisfy the Poisson bracket relationships *and* generate the correct Hamiltonian operator by substitution in the classical expression. These considerations show that, however suggestive the Ehrenfest relationships are, they are just that — a suggestive and heuristic analogy rather than a general theorem. Also, it may be a little unsporting to point out that the existence of a potential function $V(q^i)$ from which the forces may be derived ensures that the mean values of all co-ordinates q^i are independent of time. The left-hand-sides of 6.5.4 are all zero if a potential function exists, ensuring that all such equations are $0 = 0$.

This last point also illustrates the different nature of the two quantities which play the rôle of q^i in Newton's law; on the left of 6.5.4 we have the classical q^i replaced by the mean value $< q^i >$, while on the right we have q^i appearing in its own right as one of the independent co-ordinates. In Schrödinger's mechanics only the mean value may depend on t. To have a truly consistent classical appearance the quantity $< \nabla V >$ should be replaced by $\nabla_{\bar{q}} V(\bar{q})$, where $\bar{q}$ is shorthand for the mean value $< q^i >$ of each co-ordinate and $\nabla_{\bar{q}}$ is the gradient operator with respect to these quantities. It is by no means clear what the conditions might be for this to hold.

6.6   A Classical Distribution

In section 3.3 we quoted the variational condition that is the dynamical law of Schrödinger' mechanics:

$$\delta \int \left(\rho H - \rho i k \frac{\partial \Psi}{\partial t} \right) dV dt = 0 \qquad (6.6.1)$$

where $\rho = |\Psi|^2$. If we leave this expression in an essentially classical form, retaining the (real) action S, an analogous variational principle would be

$$\delta \int \left(\rho H(q^i, \nabla S, t) + \rho \frac{\partial S}{\partial t} \right) dV dt = 0 \qquad (6.6.2)$$

If this variational problem is now regarded as defining optimum ρ (the distribution of classical trajectories) and S (the momenta of those trajectories) — two *real* functions — then applcation of the standard variational calculus generates two *linked* differential equations for the functions ρ and S:

$$H(q^i, \nabla S, t) + \frac{\partial S}{\partial t} = 0 \qquad (6.6.3)$$

$$\frac{\partial \rho}{\partial t} + \nabla \cdot (\rho \vec{U}) = 0 \qquad (6.6.4)$$

where $\vec{U}$ is a vector function with components

$$U_i = \frac{\partial H}{\partial (\nabla_i S)}$$

The first equation is the Hamilton-Jacobi equation and the second is a "continuity"equation for the distribution of trajectories.

Equations similar to these may be derived by making a separation of the function Ψ into two real functions in various ways and are the basis of some semi-classical approximation schemes; one such choice is

$$\Psi = \rho^{\frac{1}{2}} e^{iS/\hbar}$$

in which the exponential form involving the imaginary action has been replaced by an alternative form $\sqrt{\rho}$. This expression, when substituted into the Schrödinger *equation* (not the Schrödinger Condition), leads to two coupled equations for the (real) functions ρ and S:

$$\frac{\partial \rho}{\partial t} + \nabla \cdot \left(\frac{\rho}{m} \nabla S \right) = 0$$

$$\frac{\partial S}{\partial t} + \frac{1}{2m} (\nabla S)^2 + V + \frac{\hbar^2}{m} \left(\frac{\nabla^2 \rho}{\rho} + \frac{(\nabla \rho)^2}{\rho^2} \right) = 0$$

The first of these equations may be interpreted as an equation of continuity for a "flux" $J = \rho \nabla S/m$ and, for a given density ρ, the second equation determines the "action". The second equation is the same as the classical Hamilton-Jacobi equation plus terms which depend on the gradients of the density; terms which we know from our earlier analysis contain contributions to the quantum-mechanical kinetic energy from the *deviations* from the means of the momenta.

Equations of this type do show how the particle distributions and their momenta are inextricably linked in quantum mechanics but are unable to cope with the full power of Schrödinger's formulation.

6.7   Velocity

The *Hamiltonian* formulation of quantum mechanics necessarily means that the description of any dynamical system has to be in terms of the position co-ordinates q^i and the momentum components p_i of the particles comprising the system. *Velocity* is relegated to a secondary rôle, defined in classical mechanics by the identity

$$v^i = \dot{q}^i = \frac{\partial H}{\partial p_i}$$

Again classically, the momentum part of the Hamiltonian is the kinetic energy term

$$T = \frac{1}{2m}|\vec{p}|^2 = \frac{1}{2m}\sum_{k,l=1}^{3N} p_k g^{kl} p_l$$

So that

$$v^k = \frac{\partial H}{\partial p_k} = \frac{\partial T}{\partial p_k} = \frac{1}{m}\sum_{l=1}^{3N} g^{kl} p_l$$

That is, apart from the factor $1/m$, the velocity components are the "index raised" components of the momentum; the contravariant components of the momentum. Naturally, in the familiar orthogonal co-ordinate systems the metric tensor is diagonal and so the velocity components are just multiples of the momentum components; in particular the familiar relationship between the Cartesian velocities and momenta is a special case.

Using this elementary result, we may re-write the "Hamiltonian" in terms of the velocity and momentum as

$$\sum_{i=1}^{3N} v^i p_i + V(q^i)$$

This quantity is not a Hamiltonian of course since it depends on velocity explicitly. We can call it R after Routh who used a function involving both velocity and momentum in a different context;[3] so

$$R = \sum_{i=1}^{3N} v^i p_i + V(q^i)$$

is bilinear in velocity and momentum components whereas the Lagrangian is quadratic in velocities and the Hamiltonian is quadratic in momenta. It should be noted, however, that the v^i and p_i are not to be regarded as *independent*; they are related by the covariant/contravariant transformation rule.

If attention is concentrated on the property of R being *linear* in momenta, we can get an insight into some of the baffling properties of Dirac's "linear" quantum mechanical equation for the electron. If any linear function of the momenta is to have the dimensions of energy, then the coefficients of these momentum components must have the dimensions (at least) of velocity and we can see from the definition of R, that these coefficients must be *interpretable* as velocities. Thus the α's of Dirac's theory play this rôle and therefore must be the velocity operators. But the α's are *algebraic* quantities not analytic ones and, in particular, have eigenvalues of $\pm c$ (the velocity of light). Dirac's linear "Hamiltonian" is, in fact, an example of our classical R. Much more important than this numerical paradox is the fact that the use of *algebraic* quantities for the velocity operators and the usual analytical quantities for the momentum components removes the covariant/contravariant relationship between momenta and velocities; the momentum components and the velocities are *independent* in the Dirac theory. It is to be expected in these circumstances that there may well be peculiarities in the velocity distributions.

We must now choose a route to define the velocity densities in the Schrödinger theory. There are two possibilities. We can simply take over the classical relationship between velocities and momenta and use our existing definition of momentum density or we can define the quantum mechanical velocity density by a relationship analogous to the classical relationship above

$$d_{v^k} = \frac{\partial d_H}{\partial d_{p^k}}$$

Since the momentum densities are complex in the Schrödinger theory we make the former choice and define the velocity density as the contravariant component of the corresponding momentum density.

---

[3] Routh did not use *both* the momentum and its associated velocity as we have done above, he used a set of linearly independent momentum and velocity components: $3N$ altogether to describe the motion of the system.

6.8   Conclusions

With a knowledge of the basic equations of Schrödinger's mechanics and the associated physical interpretation of the formalism in terms of ensembles and the probabilities generated from the distribution functions referring to these ensembles it is possible to generate other, less comprehensive, formulations of quantum mechanics and to develop certain heuristic analogies between classical and quantum mechanics. In particular, it is possible to see the relationship of the abstract, algebraic forms to the richer, interpreted mechanics of Schrödinger and to avoid some of the interpretational pitfalls which arise when only an abstract, uninterpreted skeleton is used.

Chapter 7

Conclusions

7.1 Introduction

This work is the first version of an attempt to set out Schrödinger's theory
of the dynamics of sub-atomic particles in the spirit of the mainstream
of theoretical physics; the spirit of Newton, Lagrange, Hamilton, Einstein
and Schrödinger. The aim has been to stress the continuity of method and
content between classical and quantum physics. That spirit of theoretical
physics is, by and large, that axiom systems can be abstracted from a
mature theory and are useful for certain formal processes but they play
almost no rôle in the creative thinking involved in developing a new theory
and only a minor role in understanding an existing theory.

The centre of any mathematically-expressed physical theory is a few (often just one) pairs of principles from which all developments, interpretations
and, *a fortiori*, formal structures flow as natural "fall out". These pairs of
principles always comprise a mathematical equation (or, more rarely, an
inequality) and a physical interpretation of the meaning of the equation
and the quantities appearing in that equation.

There is just one such pair of principles in Schrödinger's theory:

- (A) The dynamical law:

$$\delta \int (d_H - d_E) dV \, dt = 0$$

  The equality of the mean values of the Hamiltonian density

$$|\psi|^2 H(q^k, -\frac{i}{\psi}\frac{\partial \psi}{\partial q^k}) = d_H$$

and the energy density

$$|\psi|^2 \left(\frac{i}{\psi} \frac{\partial \psi}{\partial t} \right) = d_E$$

- (B) The physical interpretation: the complex "action" function

$$S' = S - iR$$

determines the spatial probability distributions of the particles via the square of its modulus and the means and deviations of the canonical momenta via the real and imaginary components of the complex "action" S' respectively.

From these two principles the whole of Schrödinger's theory flows without intuitive difficulty or paradox.

Now, in saying that the whole of the quantum mechanics of particles "flows from" these principles what is meant is that the development of concepts and the use of mathematical structures is "natural" even obvious. This statement is made in the context of physical theory not axiomatics. For example, the idea of "state" cannot be *logically* derived from these principles but it flows from the application of the dynamical law (A) to certain classes of physical system. The fact that "state" plays a more prominent role in quantum theory than in classical physics is not due to a difference in the logical status of "state" in the two theories but to its difference in utility and intuitive appeal in the two theories: the concept of state is *secondary* to the dynamical law. It is the fact that the particular form (A) of the dynamical law (in the particular case of conservative potentials) generates distributions of particles to which it is useful and illuminating to apply the term "state" which makes the term important in quantum theory. However true it might be that one may choose to view (A) as determining the evolution of pre-defined states of particle ensembles it is not this logical fact which gives the idea of "quantum state" its importance; it is the dynamical law which is primary.

Whatever mathematical and logical structures one may abstract from Schrödinger's theory it must be remembered that all these structures *are* abstractions from the interpreted physical theory and therefore reflect only a part of that theory; an aspect of the *structure* of the theory. The theory is not a representation of these formal structures.

It is useful to highlight some differences which arise between the formal approach and the physical approach advocated here. The most important distinction to be made in any mathematically-expressed *physical* theory is between equations and identities; a difference which is crucial to the physics but of little importance in a formal theory where relationships are either true or false and not contingently so. Indeed, it is a standard mathematical technique to transform equations into identities by limiting their domain: in differential geometry, identities in an $(n - 1)$ dimensional manifold are equations in the n dimensional enclosing space. Ever since the writings of Mach in the nineteenth century there have been attempts to present physical laws as identities or as definitions. Mach tried to present Newton's *law* $F = ma$ as a *definition* of force not as an equation relating different phyical quantities. This tendency shows itself outside of mathematical physics, for example in the pressure to *define* "survival" in terms of "fittest" or vice versa in biology. If the laws of nature are merely well-chosen definitions then they can be spun out of our heads without recourse to observation and experiment — a view which does have some prominent support[1] but which is not the view adopted here.

In quantum theory the lack of a proper distinction between identities and equations has been instrumental in the production of a large apparatus of formal theory which is almost completely empty of physical content: the use of linear operators to represent arbitrary dynamical variables.

7.2   Linear Operators and Dynamical Variables

The representation of dynamical variables by linear operators is often claimed to be central to quantum mechanics and to Schrödinger's theory in particular. This claim is made on the basis of the properties of the time-independent Schrödinger equation, of one component of this Schrödinger equation in a central field of force and of some identities involving canonical momentum components. There are several matters at issue here; let us clear up the simplest first.

The Schrödinger equation is the local condition which (together with some boundary requirements) is equivalent to the global condition (A) and embodies the dynamical law of quantum mechanics. *Only those systems occur in nature which satisfy the Schrödinger equation* (and the boundary conditions). That is, for conservative potentials only those ψ which are

---

[1] "An intelligence, unacquainted with our universe, but acquainted with the system of thought by which the human mind interprets to itself the contents of its sensory experience, should be able to attain all the knowledge of physics that we have obtained by experiment" A. Eddington 'Relativity Theory of Protons and Electrons'(Cambridge University Press) 1936 p.327. This view is made explicit by Eddington in his characteristic forthright way, but it is present as an undertone in much modern mathematical thinking.

solutions of
$$\hat{H}\psi = E\psi$$

correspond to reality; just as only those trajectories for which $F = ma$ occur in the region of applicability of classical mechanics. The question of whether or not an ensemble described by a particular ψ is capable of being realised in nature is, then, decided by the simple requirement that, for some field of force, there must be a Schrödinger equation of which that ψ is a solution (or a component of a solution in the sense of separation of variables).

Now the Schrödinger equation is a linear partial differential equation. There is also a whole family of linear differential equations and associated linear operators which arise from the *definition* of certain types of homogeneous ensemble which may or may not be realised in nature. The condition that a (single particle) ensemble be homogeneous in a canonical momentum — the momentum per particle be constant — is

$$-\frac{i}{\psi}\frac{\partial\psi}{\partial q^k} = \lambda$$

i.e.
$$\hat{P}_k\psi = \lambda\psi \tag{7.2.2}$$

a linear eigenvalue equation of very simple form but reminiscent of the time-independent Schrödinger equation in general structure. The relationships 7.2.2 are definitions i.e. *identities* (one for each k) and do not have any bearing on the satisfaction of any dynamical law; in particular (A) may or may not be satisfied by such ψ. That is, there may or may not be ensembles in nature which are represented by ψ satisfying 7.2.2.

Now in certain very special (highly symmetrical) fields of force, separation of the Schrödinger equation may yield an *equation*

$$\frac{d^2\psi}{dq^{k^2}} = a\psi \tag{7.2.3}$$

which is one component of the kinetic energy when the potential is independent of q^k. Relationships 7.2.2 and 7.2.3 have some solutions in common but they are not the same equation; they do not have the same physical interpretation and, most important, they arise in completely different ways in the theory: 7.2.2 is an identity and 7.2.3 is (a component of) an equation.

As a matter of contingent fact, ensembles of constant canonical momentum per particle often do satisfy the dynamical law — it is familiar in classical mechanics as Newtons first law in Cartesian co-ordinates.

It cannot be stressed too strongly that the Hamiltonian density (and therefore, derivatively, the "Hamiltonian operator") have a unique significance in Schrödinger's theory. It is the equality (in the mean) of the Hamiltonian density and the energy density — two independently-defined physical quantities — which *is* the dynamical law and which generates the Schrödinger equation and hence, under special circumstances, the time-independent Schrödinger equation: the eigenvalue equation. In the case of a general dynamical variable which is classically $A(q^k, p_k)$, the quantum mechanical density is

$$|\psi|^2 A(q^k, -\frac{i}{\psi}\frac{\partial \psi}{\partial q^k})$$

but there is no corresponding dynamical law; that is, there is no independently-defined dynamical quantity to which the A-density is required to be equal (in the mean or otherwise). Thus there is no variational problem to be solved for A, no time-dependent equation for A and, *a fortiori*, no time-independent *eigenvalue equation* involving a putative linear operator $\hat{A}$ "representing" A.

The trivial exceptions to this general rule are, of course, those variables A which are components of or multiples of or powers of a Hamiltonian density; most familiarly the angular kinetic energy for centrally-symmetric potentials; "the square of the angular momentum".

The difficulties and paradoxes associated with the attempt to associate a linear differential operator with each dynamical variable are well-known and cannot be solved without artifacts and *ad hoc* procedures. Indeed one can quite easily prove (see Appendix C) that such an association is, in general, impossible for variables of quadratic or higher powers in the canonical momenta where the canonical momenta are also required to satisfy the Poisson bracket conditions (the Heisenberg relationships).

But equally there is no doubt that the correct Schrödinger equation may be "deduced" for some special cases by the recipe of replacing the classical p_k by $-i\partial/\partial q^k$ in the classical Hamiltonian function; by-passing the Hamiltonian density and the variational dynamical law! The most notable cases are when the Hamiltonian is expressed in Cartesian co-ordinates and, apparently, the two co-ordinates θ, ϕ of the spherical polar system. In the absence of a vector potential the straight-forward Cartesian substitution always gives the correct Schrödinger equation and only the most innocent artifact is required to generate the correct equation in the presence of a vector potential. In the case of the square of the angular momentum apparently we also have just that; the operator representing the square of the angular momentum is just the sum of the squares of the components of the angular momentum operators.

In the case of the Hamiltonian operator obtained by substitution in Cartesian co-ordinates we are clearly dealing with a co-incidence; the very simplicity of the metric matrix in Cartesians means that

$$dV = 1 \times dQ \quad \sqrt{g} = 1$$

and the Laplacian operator "∇ squared" is indeed the square of ∇:

$$\Delta = \nabla^2 = (\nabla)^2$$

thus by-passing the action of the gradient operator on the Jacobian — colloquially "grad" and "div" are the same in Cartesians:

$$\frac{1}{h_1}\frac{\partial f}{\partial q^1} = \frac{1}{h_1 h_2 h_3}\frac{\partial}{\partial q^1}(h_2 h_3)f$$

This is only true in Cartesian co-ordinates.

Having by-passed the true dynamical law of Schrödingers mechanics by this trick, the question of what *is* the law of quantum dynamics is raised. The Schrödinger *equation* must be taken as the true dynamical law in these circumstances with the boundary conditions being plucked from the air as appropriate.

The case of the angular kinetic energy operator ("angular momentum squared") is more interesting and more illuminating because, like the *components* of angular momentum, it has been a source of confusion and paradox in the interpretation of Schrödinger's theory. We have seen that the key to understanding the paradoxes involving angular momenta is to realise that *the components of angular momentum are not canonical in the sense of Hamilton*; and, in particular, are not canonically conjugate to angular variables.

Now in the case of the "square of the angular momentum" there is a similar confusion. The square of the angular momentum is not obtained by squaring the operators conjugate to θ and ϕ but by squaring the operators representing the *Cartesian* components of the angular momentum vector. The confusion arises because, for reasons of manipulative compactness, these Cartesian components are expressed in terms of spherical polar co-ordinates:

$$\hat{\ell}_x = i\left(\cos\phi\cot\theta\frac{\partial}{\partial\phi} + \sin\phi\frac{\partial}{\partial\theta}\right)$$

$$\hat{\ell}_y = i \left(\sin\phi \cot\theta \frac{\partial}{\partial\phi} - \cos\phi \frac{\partial}{\partial\theta} \right)$$

$$\hat{\ell}_z = -i \frac{\partial}{\partial\phi}$$

Thus, the spherical polars are irrelevant to the underlying structure of the operators, the essence of the whole procedure is that one is operating in Cartesians. What is more, the entire procedure has nothing to do with the Hamiltonian canonical structure since the components of angular momentum are not canonical. Whatever its heuristic value, from the point of view of both Hamilton's and Schrödinger's theories, the formation of the "square of the angular momentum" is but a circuitous verification of the fact that, in Cartesians,

$$\Delta = \nabla^2 = (\nabla)^2$$

as before.

One could take the view that, since the Hamiltonian operator is the most important operator in quantum mechanics, and since it may be infallibly obtained in Cartesians by the substitution process and since it is a scalar and may be therefore *transformed* to arbitrary co-ordinates by standard techniques, then the Schrödinger equation may always be obtained without the use of the law (A) and its associated variational method. As we noted above this leaves the boundary conditions to be imposed arbitrarily but, more important, this essentially empirical procedure leaves the all-important question of "what *is* the dynamical law in quantum theory" unanswered. That is, no *theory* of the mechanics of sub-molecular particles is provided, just an empirical procedure for setting up certain correct equations to solve, and we are in the dark about the *meaning* of the theory and the quantities involved in the theory.

Because:

- The Hamiltonian density is not

$$\psi^* \hat{H} \psi \quad but \quad |\psi|^2 H\!\left(q^k, -\frac{i}{\psi}\frac{\partial\psi}{\partial q^k}\right)$$

and so the physical interpretation of the various densities is marred by paradox.

- Most important, the actual physical law (A) is masked by the empirical technique; quantum mechanics appears as a kind of defective adjunct to classical mechanics:

$$\hat{H}\psi = E\psi$$

is interpreted as an arbitrary extension of the classical law $H = E$ under the same conditions. Thus ideas of probability densities, the role of means and deviations, etc. must be "grafted onto"the formal structure.

As we have repeatedly hinted above, it is this latter effect which is most important if the "linear operator"technique is used in place of the dynamical law (A). There is no coherent *theory* of sub-molecular matter lying behind the *ad hoc* quantum rules and this gives the whole of quantum theory a desultory empirical air: a tendency to use a working formalism without an examination of its theoretical content. It is as if one could be content with Kepler's laws when Newton's theory is available or with the periodic classification of the elements without the underlying theory of atomic electronic structure. One is irresistibly reminded of Ptolomaic astronomy; the expansion of an unknown function in terms of known periodic functions — epicycles.

If no coherent interpreted theory of the dynamics of submolecular matter is available then the relationship between quantum and classical mechanics cannot be established. For example, in what sense can the statement "quantum mechanics goes over into classical mechanics in the limit of very high eigenvalues"be interpreted? The classical orbits of the hydrogen atom (of maximum angular momentum) are *circles* while the high energy limit solutions of the Schrödinger equation under the same conditions are three-dimensional *functions* as are all the solutions. There can be no simple relationships here in the absence of the understanding that the referents of Schrödinger's theory are *ensembles* and so there can be no one-to-one correspondence with classical mechanics whose referents are *individuals*. These considerations are quite independent of the size of Planck's constant, of course. The correct correspondence can only be made via the Hamilton-Jacobi equation: classical trajectories solve it point-by-point and quantum trajectories only in the mean, that is via the law (A).

The empirical prescription for the generation of the correct Hamiltonian operator by the substitution $-i\partial/\partial q^k$ for p_k in the classical Hamiltonian function only works in Cartesian co-ordinates and only works because, in Cartesians,

$$dV = dQ = dq^1 dq^2 dq^3$$

leading to a particularly simple form for the variational problem. One can scarcely base a whole theory of the representation of dynamical variables by linear operators on this fortuitous co-incidence and, at the same time, obscure both the main dynamical law of quantum theory and its physical interpretation.

> It is an unwritten law in physics that there shall be no priviledged co-ordinate systems.

7.3 Omissions

This work has been concerned with the general principles of Schrödinger's theory and no attempt has been made to give applications to particular systems. However, in addition to this obvious omission there are whole areas of the theory of the dynamics of sub-molecular matter which have scarcely been mentioned.

The question of the "spin"of particles has not been touched. Fields have been treated merely as sources of potential in which particles move and not as extended substances with structure and dynamics of their own and subject, presumably, to Schrödinger's dynamics. Most important of all, and to some extent including the previous omissions, the all-pervasive influence of Einstein's special theory of relativity has played no part in our deliberations. But the title of Einstein's pioneering paper was "On the Electrodynamics of Moving Bodies"and this is *precisely* what a theory of the electronic structure of matter is concerned to elucidate!

Schrödinger's mechanics depends heavily on the concepts and methods of the Hamilton-Jacobi approach to classical mechanics and the introduction of parametric methods in the Hamilton-Jacobi approach to classical mechanics introduces new problems independently of the ideas of special relativity. The introduction of a parameter on which *both* co-ordinates and time depend (a particle's "proper time"is an obvious candidate in special relativity) causes the variational problem to become the so-called "homogeneous case"and some *equations* of the variational problem are apparently satisfied *identically* or even trivially. This leads to special problems when relativity is introduced; problems which have traditionally been sidestepped either by not using the Hamiltonian formulation of relativistic mechanics or by using a freedom in the definition of the Lagrangian to choose a suitable form.

However, the Hamilton-Jacobi equation can be given a physically meaningful formulation in the homogeneous case and therefore it will prove possible to cast the dynamical law of relativistic quantum mechanics into a form analogous to (A) and it is hoped that the physical interpretation of the solutions of the equations generated from (A) will be forthcoming. These matters will be pursued elsewhere.

Appendix A

The Classical Variation Principle

In Chapter 2 Newton's equation $F = ma$ was generalised under certain fairly mild conditions to generate Lagrange's Equations which have the advantage of being valid in any arbitrary co-ordinate system. Intuitions about components of the vectors "force" and "acceleration" may fail us in unfamiliar co-ordinate systems but, if we have the scalar invariant kinetic and potential energy functions in a familiar co-ordinate frame we can generate Lagrange's Equations in any other frame by purely manipulative processes.

Again, under fairly mild constraints, it was found to be possible to express the content of Newton's Law in a terms of a single scalar function L (the Lagrangian function):

$$\frac{d}{dt}\left(\frac{\partial L}{\partial \dot{q}^i}\right) - \left(\frac{\partial L}{\partial q^i}\right) = 0 \qquad (A.1)$$

where L is a function of the general o-ordinates q^i and associated velocities $\dot{q}^i$ (and, possibly, t):

$$L = L(q^i, \dot{q}^i, t) \qquad (A.2)$$

Now, as we remarked in Chapter 2, A.1 is not a partial differential equation but is a *generator* of $3N$ ordinary differential equations: one for each q^i (and $\dot{q}^i$). The solution of A.1 is the set of q^i as functions of t: $q^i(t)$ and hence $\dot{q}^i(t)$. That is to say, treating L and A.1 as a generator of ordinary differential equations means that we have really only use for L for those q^i and $\dot{q}^i$ *which solve* A.1 and, knowing these q^i and $\dot{q}^i$ we can recover

$$\ell(t) = L(q^i(t), \dot{q}^i(t), t) \quad \text{(say)}$$

as a function of t only by simply making the substitutions.

When we use the Lagrangian function in a variational context we make an important change of viewpoint and of emphasis, which in classical mechanics is largely a mathematical device but in Schrödinger's mechanics assumes a physical significance. We treat the function L as a function of $3N + 1$ *independent* variables:

$$q^1, q^2, \ldots, q^{3N}, t$$

(treating the $\dot{q}^i$ as capable of being derived from the q^i by straightforward differentiation). That is, we regard L as having a value at each and every point in configuration space at every time not just as having values at points along the allowed trajectories $q^i(t)$.

Thus, taking two fixed points in configuration space P_1 and P_2 (say) then there are an infinity of paths between P_1 and P_2 subject only to the constraint that they do, indeed, pass through P_1 and P_2 at t_1 and t_2 (say): $q^i(t_1)$ and $\dot{q}^i(t_2)$ are fixed Clearly the integral

$$I = \int_{P_1}^{P_2} L(q^i, \dot{q}^i, t)dt \qquad (A.3)$$

is well-defined but, of course, not unique; depending on the particular path in configuration space (C, say) from P_1 to P_2. Also it is quite clear that, for non-singular transformations of co-ordinate system, L is a scalar and so I is an invariant - independent of co-ordinate frame.

If we now choose to view the values of I (which depend on C) as a *functional* of C i.e. as a functional of the particular paths taken between P_1 and P_2 :

$$I = \mathcal{L}[q^i] = \int_{P_1}^{P_2} L(q^i, \dot{q}^i, t)dt \qquad (A.4)$$

then the variational principle of mechanics states that:

> Of all possible paths C between P_1 and P_2, the one actually required by Newton's Laws - the one occurring in nature - is the one for which I is a minimum.

Where the minimum is, of course, with respect to possible paths C in configuration space i.e. possible q^i.

Clearly, this condition must, if it is to be useful, generate differential equations for the q^i since one can scarcely hope to find the optimal q^i by exhaustive trial and error; by substitution in A.4 and searching for a minimum. These differential equations can be found by elementary means.

Let us assume that we have a one-parameter family of curves "near"to the optimising curve C and let the family be distinguished one from another by the parameter λ

$$q^i(t;\lambda) \quad i = 1,3N \tag{A.5}$$

and we may, without loss of generality, assume that the minimising curve C corresponds to $\lambda = 0$. Now, if all the curves in the family *are* sufficiently close to C, we may neglect terms of order greater than one in the Taylor expansion in λ and write

$$q^i(t;\lambda) = q^i(t;0) + \delta q^i(t) \tag{A.6}$$

where

$$\delta q^i(t) = \lambda \left(\frac{\partial q^i}{\partial \lambda}\right)_{\lambda=0} = \lambda \eta^i(t) \quad \text{(say)}$$

We are also insisting that *all* the curves must pass through P_1 and P_2 so that

$$\delta q^i(t_1) = \delta q^i(t_2) = 0$$

In order to investigate the behaviour of A.4 with respect to changes in the $\dot{q}^i$ we need

$$\dot{q}^i(t;\lambda) = \dot{q}^i(t;0) + \frac{d}{dt}\left(\delta q^i(t)\right)$$

$$= \dot{q}^i(t;0) + \lambda\frac{d}{dt}\left(\frac{\partial q^i}{\partial \lambda}\right)_{\lambda=0}$$

$$= \dot{q}^i(t;0) + \lambda\dot{\eta}^i$$

i.e.

$$\dot{q}^i(t;\lambda) = \dot{q}^i(t;0) + \delta\dot{q}^i(t) \tag{A.7}$$

(the δ "operator"and differentiation "commute"). Now, of course, we may view I (a functional of the q^i) as a *function* (I) of the parameter λ and we have

$$I(\lambda) = \int_{t_1}^{t_2} L(q^i(t;\lambda), \dot{q}^i(t;\lambda), t)dt$$

$$= \int_{t_1}^{t_2} L(q^i(t) + \lambda\eta^i(t), \dot{q}^i(t) + \lambda\dot{\eta}^i(t), t)dt$$

The sought-after optimising C is, of course, $I(0)$:

$$I(0) = \int_{t_1}^{t_2} L(q^i(t), \dot{q}^i(t), t)dt$$

so that

$$I(\lambda) = I(0) + \lambda \left(\frac{\partial I}{\partial \lambda}\right)_{\lambda=0} + \text{ terms of order } \lambda^2$$

We therefore define

$$\delta I = \left(\frac{\partial I}{\partial \lambda}\right)_{\lambda=0} \qquad (A.8)$$

Thus, the condition that $I(\lambda)$ have a (local) minimum is then

$$\delta I = 0 \qquad (A.9)$$

at least.

Now

$$\delta I = \left(\frac{\partial I}{\partial \lambda}\right)_{\lambda=0} = \int_{t_1}^{t_2} \sum_{i=1}^{3N} \left(\frac{\partial L}{\partial q^i}\eta^i + \frac{\partial L}{\partial \dot{q}^i}\dot{\eta}^i\right) dt \qquad (A.10)$$

If this integral is to be zero for *all* the possible family η^i and to generate a differential equation for the optimising trajectories $q^i(t)$, we must cast this expression into a form in which the arbitary functions η^i appear as a *factor* in the integrand. As presently constituted, both η^i and $\dot{\eta}^i$ appear in the integrand. Thus, we integrate the second set of terms by parts; a typical term being

$$\int_{t_1}^{t_2} \frac{\partial L}{\partial \dot{q}^i}\dot{\eta}^i\, dt = \left[\eta^i \frac{\partial L}{\partial \dot{q}^i}\right]_{t_1}^{t_2} - \int_{t_1}^{t_2} \eta_i \frac{d}{dt}\left(\frac{\partial L}{\partial \dot{q}^i}\right) dt$$

The first term is zero because

$$\eta^i(t_1) = \eta^i(t_2) = 0 \quad (\text{ all i})$$

so that

$$\delta I = \int_{t_1}^{t_2} \sum_{i=1}^{3N} \left(\frac{\partial L}{\partial q^i} - \frac{d}{dt}\left(\frac{\partial L}{\partial \dot{q}^i}\right)\right) \eta^i\, dt = 0$$

this is true for arbitrary η^i if each of the factors of η^i in the integrand is zero:

$$\frac{\partial L}{\partial q^i} - \frac{d}{dt}\left(\frac{\partial L}{\partial \dot{q}^i}\right) = 0 \qquad (A.11)$$

which are precisely the Lagrange equations.

Another set of equations may be derived from A.10 by reducing the integrand to a form which only depends on $\dot{\eta}^i$ (i.e. not on η^i). This may be done by integrating the *first* set of terms in A.10 by parts to obtain

$$\int_{t_1}^{t_2} \frac{\partial L}{\partial q^i} \eta^i\, dt = \left[\eta^i \int_{t_1}^{t_2} \frac{\partial L}{\partial q^i} dt\right]_{t_1}^{t_2}$$
$$-\int_{t_1}^{t_2} \left(\int_{t_1}^{t_2} \frac{\partial L}{\partial q^i} dt\right) \dot{\eta}^i\, dt$$

again, the first term vanishes because of the boundary conditions so that A.10 becomes

$$\delta I = \int_{t_1}^{t_2} \left(\frac{\partial L}{\partial \dot{q}^i} - \int_{t_1}^{t_2} \left(\frac{\partial L}{\partial q^i}\right) dt \dot{\eta}^i\right) dt$$

This time, however, we do not have a straightforward condition on $\dot{\eta}^i$ to obtain an equation determining the allowed trajectories directly. However, it can be shown that $\delta I = 0$ is equivalent to

$$\frac{\partial L}{\partial \dot{q}^i} - \int_{t_1}^{t_2} \frac{\partial L}{\partial q^i} dt = \quad \text{constant} \qquad (A.12)$$

(for all i) which, on differentiation with respect to t, yields

$$\frac{d}{dt}\left(\frac{\partial L}{\partial \dot{q}^i}\right) - \frac{\partial L}{\partial q^i} = 0$$

Lagrange's equations, as before. Thus, considerations of variations in the q^i and the associated $\dot{q}^i$ lead to a consistent set of equations which determine C, the optimum path in configuration space between the points P_1 and P_2; that is, in ordinary three-dimensional space, the dynamically allowed trajectories of the particles of the system, the $q^i(t)$ ($i = 1, 3N$).

   To show that the extremum we have determined is actually a minimum requires taking the analysis to second order in λ, defining an appropriate "second variation" $\delta^2 I$ and generating the condition that

$$\delta^2 I \geq 0$$

In point of fact, the generation of the Lagrange equations is sufficient for our purposes.

   These considerations do not, of course, take us one step further than our original generalisation of Newton's Law from the point of view of that actual computation of particle trajectories. They do, however, consolidate the concept that the Lagrangian function is defined for all values of the q^i and $\dot{q}^i$ not simply the ones which solve Lagrange's equations. In this view the variation principle *picks out* from all possible q^i, having the same status as t as independent variables, the $q^i(t)$ which *embody the physical law* of classical mechanics.

This view of the q^i and $\dot{q}^i$ enables the generalised momenta of Hamilton to be introduced *as definitions* on the same footing as the q^i, independent variables which *on satisfaction of a physical law*, become functions of t: $p^i(t)$. The Hamiltonian *function* is a function of the q^i and p_i (plus, possibly, t) but when the dynamical law is imposed and the q^i, p_i become *dependent* on t: $q^i(t), p_i(t)$ then the Hamiltonian is equal to the energy of the system.

With hindsight it is obvious that the derivation of A.11 from $\delta I = 0$ is unaffected by the addition of a total time derivative to the integrand L:

$$\delta \int \left(L + \frac{dF}{dt} \right) dt = 0$$

also generates equations A.11 if variations in the original integral do. This is not the source of any mathematical or conceptual difficulty since we always know L as an explicit function of the q^i, $\dot{q}^i$ and t. Only if we try to *infer* a Lagrangian from the assumed known equations of motion (i.e. *reverse* the arguments given in this Appendix) will the reasoning fail. In fact, this apparent arbitrariness in the definition of the Lagrangian will be given a physical meaning in the transformation theory.

Appendix B

Vector Potentials

The general form of the classical Hamiltonian function for a particle in the presence of vector and scalar potentials is:

$$\hat{H} = \frac{1}{2m} \sum_{k,l=1}^{3N} p_k' g^{kl} p_l' + V$$

where

$$p_k' = p_k + aA_k = p_k + b_k \quad (say)$$

where A_k is a component of the vector potential, p_k is a canonical momentum, $V = V(q^1, q^2, \ldots)$ is the potential energy expression due to the scalar potential and a is a constant depending on the units employed.

Thus, in Schrödinger's theory the Hamiltonian density is, using

$$p_k = -\frac{i}{\psi} \frac{\partial \psi}{\partial q^k}$$

$$d_H = \frac{1}{2m} \sum_{k,l=1}^{3N} |\psi|^2 g^{kl} \left(\frac{i}{\psi^*} \frac{\partial \psi^*}{\partial q^k} + b_k^* \psi^* \right) \left(-\frac{i}{\psi} \frac{\partial \psi}{\partial q^l} + b_l^* \psi \right) + |\psi|^2 V(q^1, q^2, \ldots)$$

which becomes, using $|\psi|^2 = \psi^* \psi$ and multiplying out:

$$d_H = \frac{1}{2m} \sum_{k,l=1}^{3N} g^{kl} \left(i \frac{\partial \psi^*}{\partial q^k} + b_k^* \psi^* \right) \left(-\frac{i}{\psi} \frac{\partial \psi}{\partial q^l} + b_l \psi \right) + |\psi|^2 V$$

or, defining new operators $\hat{D}_k$,

$$d_H = \frac{1}{2m} \sum_{k,l=1}^{3N} g^{kl} \left(\hat{D}_k^* \psi^* \right) \left(\hat{D}_l \psi \right) + V |\psi|^2$$

$$= \frac{1}{2m}|\hat{D}\psi|^2 + V|\psi|^2$$

where

$$\hat{D}_k = (-i\frac{\partial}{\partial q^k} + b_k)\vec{e}_k$$

and

$$\hat{D}\psi = \sum_{k=1}^{3N} \vec{e}_k \hat{D}_k \psi$$

where the $\vec{e}_k$ are the natural unit tangent vectors of the co-ordinate system (q^1, q^2, q^3) .

With this expression for d_H the dynamical law is, as usual

$$\delta \int (d_H - d_E)dVdt = 0$$

Rather than carry through the detailed derivation of the Schrödinger equation for this special case it is as easy to do the derivation in general for the minimisation of a general functional depending on a function ϕ , its conjugate ϕ^* and the quantities $\hat{D}_k\phi$ where

$$\hat{D}_k\phi = -i\frac{\partial\phi}{\partial q^k} + b_k\phi = -i\partial_k\phi + b_k\phi$$

where the b_k are (fixed) real functions of space and time since the vector potential components are real.

Therefore let

$$\mathcal{B} = \int B(\phi, \hat{D}_k\phi, \phi^*, \hat{D}_k\phi^*, \partial_t\phi, \partial_t\phi^*)dVdt$$

be minimised with respect to variations in

$$\phi, \phi^*, \hat{D}_k\phi, \hat{D}_k\phi^*, \partial_t\phi, \partial_t\phi^*$$

as before. It is only necessary to sketch the differences between this case and the case discussed in Section 3.5; that is the form of the *spatial* dependence of B on ϕ, ϕ^*, etc.

As before, we assume the validity of the Taylor expansion about the actual minimising ϕ and obtain

$$\delta\mathcal{B} = \int \delta B dVdt$$

where

$$\delta B = \frac{\partial B}{\partial \phi}\delta\phi + \sum_{j=1}^{3N}\frac{\partial B}{\partial(\hat{D}_j\phi)}\delta(\hat{D}_j\phi) + \quad \text{etc.}$$

and, with $\delta\phi = \epsilon\eta$ for "small" ϵ

$$\delta B = \epsilon\frac{\partial B}{\partial \phi}\eta + \epsilon\sum_{j=1}^{3N}\frac{\partial B}{\partial(\hat{D}_j\phi)}\hat{D}_j\eta + \dots$$

Now

$$\delta B = \epsilon\int \eta\left(\frac{\partial B}{\partial \phi}\eta + \sum_{j=1}^{3N}\frac{\partial B}{\partial(\hat{D}_j\phi)}\hat{D}_j\eta + \dots\right) dVdt$$

where $dV = \sqrt{g}dQ = \sqrt{g}dq^1dq^2dq^3$

The second set of terms has a typical member

$$\int \frac{\partial B}{\partial(\hat{D}_j\phi)}\hat{D}_j\eta\sqrt{g}dQdt = \int \frac{\partial B}{\partial(\hat{D}_j\phi)}\left(-i\frac{\partial\eta}{\partial q^j} + b_k\eta\right)\sqrt{g}dQdt$$

and the first term may be integrated by parts exactly as before to yield a boundary term and a term containing η as a factor; giving, for the whole typical term,

$$\int\left\{i\partial_j\left(\sqrt{g}\frac{\partial B}{\partial(\hat{D}_j\phi)}\right) + b_j\frac{\partial B}{\partial(\hat{D}_j\phi)}\right\}\eta dQdt$$

$$= \int \eta\left\{\frac{i}{\sqrt{g}}\partial_j\left(\sqrt{g}\frac{\partial B}{\partial(\hat{D}_j\phi)}\right) + b_k\frac{\partial B}{\partial(\hat{D}_j\phi)}\right\} dVdt$$

$$= \int \eta\left\{\frac{1}{\sqrt{g}}\hat{D}_j^*\left(\sqrt{g}\frac{\partial B}{\partial(\hat{D}_j\phi)}\right)\right\} dVdt$$

where $\hat{D}_j^*f = i\partial_jf + b_jf$

Thus, the spatial part of the variation yields

$$\delta B = \epsilon\int \eta\left\{\frac{\partial B}{\partial \phi} + \sum \frac{1}{\sqrt{g}}\hat{D}_j^*\left(\sqrt{g}\frac{\partial B}{\partial(\hat{D}_j\phi)}\right)\right\} dVdt$$

to first order in ϵ.

It is readily seen that, comparing this result with the result of Section 3.5, simply repacing ∂_j by $\hat{D}_j$ gives a result of identical form. Simple substitution then generates the Schrödinger equation by replacing B by d_H.

Appendix C

Momentum "Operators"

The momentum operator in both the Hamilton-Jacobi and in Schrödinger's theory is $\partial/\partial q^k$; the difference between the two theories is not the form of the momentum operator but the class of function on which it operates and the dynamical law which defines allowed trajectories. However, it is still occasionally claimed that the Hamiltonian operater of Schrödinger's theory can be generated by the substitution of suitably-defined momentum operators $\hat{p}_k$ for the classical momentum components in the expression for the Hamiltonian function of classical mechanics:

$$H = \frac{1}{2} \sum_{k,l=1}^{3N} p_k g^{kl} p_l + V(q^k) \qquad \text{(C.13)}$$

Now it is known from the derivation of the Schrödinger Equation from the Schrödinger Condition that the quantum mechanical Hamiltonian operator contains the Laplacian operator ∇^2:

$$\hat{H} = \frac{1}{2m} \sum_{k,l=1}^{3N} -\frac{1}{\sqrt{g}} \frac{\partial}{\partial q^k} \sqrt{g} g^{kl} \frac{\partial}{\partial q^l} + V(q^j) \qquad \text{(C.14)}$$

where

$$g = det|g^{kl}|$$

and, simply because $g^{kl} = \delta^{kl}$ in Cartesians, this transformation is clearly possible for Cartesian co-ordinates. However, the unsymmetrical form way in which the partial derivatives appear in C.14 does not hold out much hope of the substitution working in the general case; it clearly depends on some fortunate co-incidences in the derivatives of the metric tensor and its determinant if it is to work in general. Moreover, it is clear by simple inspection that the simple form $\partial/\partial q^k$ can never generate the correct Hamiltonian; a new form of the momentum operator is needed at the very least if we are to have any hope of success.

Now the postulation of a new form for the operator which is central to the theory *without good scientific reason* (i.e. simply on mathematical grounds) is the worst kind of hostage to fortune, but let us continue. Let us assume that the new operator is linear in $\partial/\partial q^k$ i.e of the general form

$$\hat{p}_k = -ia_k \frac{\partial}{\partial q^k} (b_k + c_k$$

where a_k, b_k, c_k are, as yet, arbitrary functions of the q^k; the left bracket indicates that, for example,

$$\hat{p}_k f(q^j) = -ia_k \frac{\partial}{\partial q^k}(b_k f) + c_k f$$

Now elementary considerations show that this form is, in fact, equivalent to a simpler form since

$$\frac{1}{b_k} \frac{\partial(b_k f)}{\partial q^k} = f \frac{\partial \ln(b_k)}{\partial q^k} + \frac{\partial f}{\partial q^k}$$

we may therefore assume a simpler form

$$\hat{p}_k = -ia_k \frac{\partial}{\partial q^k} + c_k \tag{C.15}$$

without loss of generality.

If we drop the potential energy term $V(q^j)$ from both expressions, the classical Hamiltonian with the above form for the p_k becomes:

$$\hat{H} = \frac{1}{2} \sum_{k,l=1}^{3N} \left(-ia_k \frac{\partial}{\partial q^k} + c_k \right) g^{kl} \left(-ia_l \frac{\partial}{\partial q^l} + c_l \right)$$

Now, if this expression is to be *identically* equal to the quantum mechanical Hamiltonian operator we may equate coefficients of $\partial/\partial q^j$, including the zero-order term and obtain expressions for the a_k and c_k. This leads to

$$a_k = 1 \quad (k = 1, 3)$$

$$c_k = \frac{1}{g} \frac{\partial g}{\partial q^k} \quad (k = 1, 3)$$

which fixes the arbitrary functions. But the dissappearance of the zero-order term (to ensure that there is no spurious "potential energy" term arising from the "kinetic energy"operator) generates another relationship which, in view of the fact that the a_k, c_k are *fixed*, must be an *equation*

$$\sum_{k,l=1}^{3N} c_k g^{kl} c_l = 0$$

i.e.

$$\frac{1}{g^2} \sum_{k,l=1}^{3N} \frac{\partial g}{\partial q^k} g^{kl} \frac{\partial g}{\partial q^l} = 0 \qquad \text{(C.16)}$$

But it is trivial to verify that this equation is not satisfied by some of the most familiar orthogonal co-ordinate systems, let alone arbitrary co-ordinates; it is not true in spherical polars, for example.

That is, it is not possible to generate the Hamiltonian operator of Schrödinger's mechanics by substitution of an operator linear in $\partial/\partial q_k$ into the expression for the Hamiltonian function of classical mechanics. It is elementary to verify that the use of any re-arranged form of the product

$$\sum_{k,l=1}^{3N} p_k g^{kl} p_l$$

e.g.

$$\sum_{k,l=1}^{3N} p_k p_l g^{kl} \quad \text{or} \quad \sum_{k,l=1}^{3N} g^{kl} p_k p_l$$

(or linear combinations of these) does not overcome the difficulty except possibly in specific cases - there is no general solution to the problem. We are therefore spared the embarassment of casting about for an interpretation of any new form of the momentum operator.

With the benefit of the interpretation of Schrödinger's mechanics via the physical significance of the Schrödinger Condition and the densities which it contains, it is obvious why such a transformation cannot be found. In classical mechanics when the dynamical law is imposed (i.e. when the Hamilton-Jacobi equation is satisfied) the Hamiltonian function has numerical values equal to the energy of the system. But the "Hamiltonian"operator of Schrödinger's mechanics does not generate the energy density even when the dynamical law (Schrödinger's Condition or the Schrödinger equation) is satisfied. This Hamiltonian operator only generates the *mean value* of the energy and so the physical interpretation of the two "Hamiltonians"and their status in their relevant theories is quite different and we are therefore relieved to find that they are mathematically unrelated except indirectly via the dynamical law - the Schrödinger Condition.

> The crux of the matter is, of course, that the kinetic energy
> density is the square of the momentum density *not* the
> density of the square of the momentum "operators".

One final point on momentum "operators"; the derivative of a function
on a closed interval is not defined at the endpoints. Thus we must make
some decision about the momentum density at the end-points of a closed
interval. This mathematical necessity is, presumably, the equivalent of
resolving the problem of "impacts" with a fixed object. One way to resolve
this problem is to insist that the momentum operator be Hermitian at
these points which generates an additional contribution to the momentum
density expressible as the addition of a $\delta-$function to the usual momentum
operator.

Appendix D

Ensembles and Abstract Objects

The general laws of science hold for "abstract"or "idealised"objects. That is, for example, celestial mechanics (at a certain level of approximation) deals with the relative motions of bodies which have the mass of the sun and of Jupiter. In quantum mechanics the referents of the theory are ensembles of identical systems. Now both of these examples can be reduced to the same type of abstract object. One could just as well say that the referents of celestial mechanics (in this example) were the set (ensemble?) of all possible pairs of bodies with the given masses (those of the sun and of Jupiter). However the processes of abstraction may be formally similar but are intuitively and conceptually different.

When we use celestial mechanics to calculate the (unperturbed!) orbit of Jupiter we have in mind the fact the result of our calculation will be an approximation to the orbit of a single concrete object which will be capable of being experimentally verified. We have, in presenting Jupiter and the sun as merely point masses, interacting under gravity, abstracted from all kinds of properties of these two bodies: their compositions, densities, colours, temperatures etc. etc. We have, in a word, *idealised* interacting bodies by abstracting from these, as we say, inessential properties. In performing this abstraction we have, from a mathematical point of view (inadvertantly, as it were) constructed an ensemble of systems: the set of all pairs of objects of these masses acting under gravity with given "initial"conditions. This latter "abstract object"is the referent of our calculations.

However, in view of the fact that we are unlikely to come across other suns with other planets with the same masses, the ensemble has a rather artificial look about it; it is of mathematical origin rather than of physical significance.

In quantum mechanics the situation is quite different. We may picture the hydrogen atom as a proton and an electron in mutual interaction under an inverse-square law of attraction, and make as close an analogy with the sun/Jupiter system as we can. This time there are two *sorts* of difference between the concrete system and the abstract system:

- There actually are, in reality, many, many identical systems of this kind existing in nature.

- We do not have the "initial conditions"for the motion or anything which can effectively take the place of those initial conditions.

What quantum mechanics does, colloquially speaking, is to make a virtue of necessity and combine these two properties and, in its definition of the referent of the theory, abstract from "initial conditions"or "position in space"and "use"the fact that there are many, many identical systems to arrive at an abstract object which, unlike the case of the sun and Jupiter, abstraction is from the dynamical properties of the particles rather than their static physical properties. Indeed, the current view is that the static properties of electrons and of protons are all the same so that one cannot form an abstract object except by abstracting from the dynamical properties of sub-atomic particles (position, velocity, angular momentum, energy . etc.). But macroscopic bodies always differ in one or more of their static properties (colour, density etc etc.), so that there are more possibilities for the formation of abstract objects.

With these considerations in mind, it is perhaps not too pedantic to emphasise the intuitively different nature of "macroscopic abstract objects"and "sub-atomic abstract objects". The question is, of course, should one retain the term "abstract object"for either or both. It seems best to retain the term to cover both cases and use a subsidiary term for each of the two sub-classes.

In the macroscopic case the process of abstraction is, in physical science, essential to get to grips with the actual application of the theory to particular objects: these objects are, as it were, *idealised* by the abstraction process; becoming, for example, point masses rather than real planets. It seems natural, therefore, to denote an abstract object obtained by abstraction from its "incidental"(from the point of view of a general theory) static, intrinsic properties as an *idealised object*. A general theory cannot normally say anything about the values of the properties which have been ignored in the formation of an idealised object. One could hope that some information might be available about the average value of such properties; but in the case of static properties this is clearly meaningless. Is one interested, for example, in the average colour of an abstract planet?

In contrast, in the microscopic case where abstraction is only possible from the dynamical properties of the concrete objects and the prediction of average values of these dynamical properties is useful and meaningful, it is proposed to denote an abstract object formed in this way by the traditional term *ensemble*.

The key difference between the two concepts is brought out by the idea of averages over the quantities which do not appear in the definition of the abstract object. In the case of an idealised object the theory being applied makes no use or mention of the properties which are abstracted from: celestial mechanics is independent of the colour and chemical composition etc. of the planets. But the position of electrons and protons *do* do appear in the quantum mechanics of the hydrogen atom. That is, those properties of the sub-atomic particles which have been abstracted away in the formation of an ensemble do appear in the theory and the theory is able to make predictions of the average values of quantities not appearing in the definition of abstract objects of the "ensemble" type.

A little thought is enough to convince that this distinction between idealised objects and ensembles is more fundamental than the distinction based on static (intrinsic) and dynamical (extrinsic) properties. If an abstract object is defined by abstraction from some property (or properties) which is, nevertheless used in the theory of that abstract object, then the theory is a probabilistic one. That is, in our example above, comparing celestial mechanics with quantum mechanics, quantum mechanics is a probabilistic theory but celestial mechanics is not. We therefore change our definition of the two types of abstract object to reflect this new point of view:

> An abstract object (formed by abstraction from properties of concrete objects) which is used in a theory which contains no mention of the properties not appearing in the definition of the abstract object is called an *idealised object*

> An abstract object which is used in a theory which contains references to the properties of the abstract object not appearing in the definition of that abstract object is called an *ensemble*.

For emphasis, probabilistic science deals with abstract objects of the type ensemble.

One may state this difference between concrete systems and the two types of abstract system colloquially as "abstract objects only have definate values of the properties mentioned in their names". Now, clearly, this is straining the use of "name" a little, it is unusual even for abstract objects to have really comprehensive names. But it is a useful idea since it emphasises the principal difference between concrete objects and abstract objects. If we compare the "names" of a classical hydrogen atom and a quantum hydrogen atom we see that they are, respectively,

> "A proton and an electron in mutual (electrostatic) interaction with initial relative positions q^k and initial relative momenta p_k."

and

> "A proton and an electron in mutual (electrostatic) interaction."

Showing how the position and momentum appear in the classical name but not in the quantum name.